A Photographic Atlas

for the

Biology Laboratory

SIXTH EDITION

Kent M. Van De Graaff
Weber State University

John L. Crawley

Morton Publishing Company
925 W. Kenyon, Unit 12
Englewood, Colorado 80110

*To our teachers, colleagues, friends, and students
who share with us a mutual love for biology.*

The tufted puffin, *Fratercula cirrhata,* is a pelagic seabird
found in the North Pacific Ocean.

Preface to Sixth Edition

Biology is an exciting, dynamic, and challenging science. It is the study of life. You are fortunate to be living at a time when insights and discoveries in all aspects of biology are occurring at a very rapid pace. Much of the knowledge that you will learn in a biology course has application in improving humanity and the quality of life. An understanding of basic biology is essential for you to establish a secure foundation for more advanced courses in the life sciences or health sciences. The principles taught in a basic biology course are of immeasurable value in dealing with the ecological and bioethical problems currently facing us.

Biology is a visually oriented science. The sixth edition of *A Photographic Atlas for the Biology Laboratory* is designed to provide you with quality photographs of organisms, similar to those you may have the opportunity to observe in a biology laboratory. It is designed to accompany any biology text or laboratory manual you may be using in the classroom. In certain courses, this photographic atlas could serve as the laboratory manual.

An objective of this atlas is to provide you with a balanced visual representation of the kingdoms of biological organisms. Great care has been taken to construct completely labeled, informative figures that are depicted clearly and accurately. Many new photographs have been added to this edition of the biology atlas. The specimens depicted appear as you see them in the biology laboratory. The updated classification and terminology used in this atlas are the same as those found in the more commonly used college biology texts.

Numerous dissections of plants and invertebrate and vertebrate animals were completed and photographed in the preparation of this atlas. These images are included for those students who have the opportunity to do similar dissections as part of their laboratory requirement. In addition, photographs of a sheep heart dissection are included in this atlas.

Chapter 9 of this atlas is devoted to the biology of the human organism, which is emphasized in many biology textbooks and courses. In this chapter, you are provided with a complete set of photographs for each of the human body systems. Human cadavers have been carefully dissected and photographed to clearly depict each of the principal organs from each of the body systems. Selected radiographs (X-rays), CT scans, and MR images depict structures from living persons and thus provide an applied dimension to this portion of the atlas.

The success of the previous editions of *A Photographic Atlas for the Biology Laboratory* provided opportunities to make changes to enhance the value of this new edition in aiding students in learning about plants and animals. The total revision of this atlas presented in its fifth edition required an inordinate amount of planning, organization, and work. As authors we have the opportunity and obligation to listen to the critiques and suggestions from students and faculty who have used this atlas. This constructive input is appreciated and has resulted in a greatly improved atlas.

One objective in preparing this edition of the atlas was to create an inviting pedagogy. The page layout was improved by careful selection of updated, new, and replacement photographs. Many illustrations were redrawn to be more appealing and accurate. Each image in this atlas was carefully evaluated for its quality, effectiveness, and accuracy. Quality photographs of detailed dissections were completed, enhancing the value of this edition.

Prelude

Scientists work to determine accuracy in understanding the relationship of organisms even when it requires changing established concepts. Development, structure, function, and the fossil record and geological dating are used to establish systematics and classify organisms. As new techniques become available, they too aid in our understanding of evolutionary relationships between groups of organisms and closely related species.

In 1758 Carolus Linnaeus, a Swedish naturalist, assigned all known kinds of organisms into two kingdoms—plants and animals. For more than two centuries, this dichotomy of plants and animals served biologists well. In 1969, Robert H. Whittaker convincingly made a case for a five-kingdom system composed of Monera, Protista, Fungi, Plantae, and Animalia. The five-kingdom basis of systematics prevailed for more than 20 years, but has been challenged with a new system that includes three domains (see exhibit 1). This new system, which is used in this edition of *A Photographic Atlas for the Botany Laboratory*, is based on criteria used in the past and new techniques in molecular biology. It is important to note, however, that a classification scheme is a human construct subject to alteration as additional knowledge is obtained.

Exhibit 1 Domains, Kingdoms, and Representative Examples

Bacteria Domain– Cyanobacteria, gram-negative and gram-positive bacteria

Archaea Domain– Methanogens, halophiles, and thermophiles

Eukarota Domain– Eukaryotes, single-celled and multicelled organisms; fungi, protists, plants, and animals

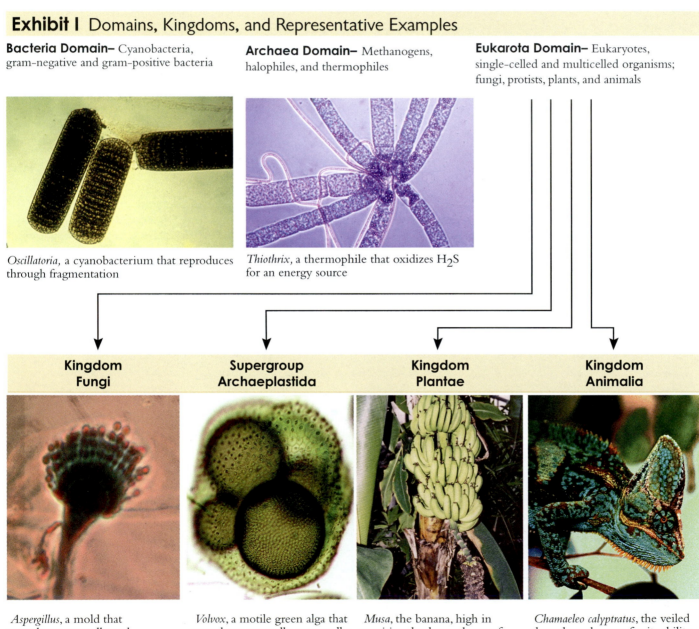

Oscillatoria, a cyanobacterium that reproduces through fragmentation

Thiothrix, a thermophile that oxidizes H_2S for an energy source

Kingdom Fungi

Supergroup Archaeplastida

Kingdom Plantae

Kingdom Animalia

Aspergillus, a mold that reproduces asexually and sometimes sexually

Volvox, a motile green alga that reproduces asexually or sexually

Musa, the banana, high in nutritional value and one of the twelve most important human food plants

Chamaeleo calyptratus, the veiled chameleon, known for its ability to change colors according to its mood

Basic Characteristics of Domains

Domain	Characteristics
Bacteria Domain—Bacteria	Prokaryotic cell; single circular chromosome; cell wall containing peptidoglycan; chemosynthetic autotrophs, chlorophyll-based photosynthesis, photosynthetic autotrophs, and heterotrophs; gram-negative and gram-positive forms; lacking nuclear envelope; lacking organelles and cytoskeleton
Archaea Domain—Archaea	Prokaryotic cell; single circular chromosome; cell wall; unique membrane lipids, ribosomes, and RNA sequences; lacking nuclear envelope; some with chlorophyll-based photosynthesis; with organelles, and cytoskeleton
Eukarota Domain—Eukarya	Single-celled and multicelled organisms; nuclear envelope enclosing more than one linear chromosome; membrane-bound organelles in most; some with chlorophyll-based photosynthesis

Common Classification System of Some groups of Living Eukaryotes

Eukaryote Supergroups

    Amoebozoa– amoeboid protozoa, cellular and plasmodial slime molds

    Rhizaria– cercozoans, foraminiferans, radiolarians, and haplosporidia

    Chromalveolata– cryptophytes, haptophytes, stramenopiles, alveolates, and diatoms

    Excavata– diplomonads, parabasalids, euglenoids, and kinetiplastids

    Archaeplastida– red algae, green algae, and land plants

    Opistokonta– fungi, choanoflagellates, and metazoa (multicellular animals)

* Kingdom Protista– heterotrophic and photosynthetic protists
- Phylum Myxomycota– plasmodial slime molds
- Phylum Dictyosteliomycota (=Acrasiomycota)– cellular slime molds
- Phylum Oomycota– water molds
- Phylum Euglenophyta– euglenoids
- Phylum Cryptophyta– cryptomonads
- Phylum Rhodophyta– red algae
- Phylum Dinophyta (=Pyrrhophyta)– dinoflagellates
- Phylum Haptophyta– haptophtes
- Phylum Bacillariophyta– diatoms (diatoms are often placed in the phylum Chrysophyta)
- Phylum Chrysophyta– chrysophytes
- Phylum Phaeophyta– brown algae
- Phylum Chlorophyta– green algae

Kingdom Fungi
- Phylum Chytridiomycota– chytrids
- Phylum Zygomycota– zygomycetes
- Phylum Glomeromycota– glomeromycetes
- Phylum Ascomycota– ascomycetes
- Phylum Basidiomycota– basidiomycetes

Kingdom Plantae– bryophytes and vascular plants
- Phylum Hepatophyta– liverworts
- Phylum Anthocerophyta– hornworts
- Phylum Bryophyta– mosses
- Phylum Lycophyta (=Lycopodiophyta)– club moss, ground pines, and spike mosses
- Phylum Pteridophyta– wisk ferns, horsetails, and ferns
- Phylum Cycadophyta– cycads
- Phylum Ginkgophyta– Ginkgo
- Phylum Pinophyta (=Coniferophyta)– conifers
- Phylum Gnetophyta– gnetophytes
- Phylum Magnoliophyta (=Anthophyta)– angiosperms (flowering plants)

** Kingdom Animalia– invertebrate and vertebrate animals
- Phylum Porifera- sponges
- Phylum Cnidaria- coral, hydra, and jellyfish
- Phylum Ctenophora- comb jellies
- Phylum Platyhelminthes- flatworms
- Phylum Rotifera- rotifers
- Phylum Bryozoa (=Ectoprocta)
- Phylum Brachiopoda
- Phylum Phoronida
- Phylum Nemertea- proboscis worms
- Phylum Mollusca- clams, snails, and squids
- Phylum Annelida- segmented worms
- Phylum Nematoda- round worms
- Phylum Arthropoda- crustaceans, insects, and spiders
- Phylum Echinodermata- sea stars and sea urchines
- Phylum Chordata- lancelets, tunicates, and vertebrates

* **Historically considered a Kingdom, protists are no longer recognized as such in modern taxonomy. For convenient reference to earlier classification schemes, protist phyla are presented here, but note that each of these are depicted more accurately within the Eukaryote Supergroups.**

** Only selective information of the Kingdoms Animalia is presented in this atlas.

Acknowledgments

Many professionals have assisted in the preparation of *A Photographic Atlas for the Biology Laboratory,* sixth edition, and have shared our enthusiasm of its value for students of biology. We are especially appreciative of Dr. Byron Adams at Brigham Young University, Donald C. Leynaud at Wabash Valley College, Mt Carmel, IL, Jennifer Clevinger at Walsh University, North Canton, OH, Ann Fraser at Kalamazoo College Kalamazoo, MI, Lori Rose at Sam Houston State University, and Terese Dudek, Kishwaukee College, Malta, IL, for their detailed review of this atlas. Drs. Ronald A. Meyers, John F. Mull, and Samuel I. Zeveloff of the Department of Zoology at Weber State University and Dr. Samuel R. Rushforth at Utah Valley University were especially helpful and supportive of this project. The radiographs, CT scans, and MR images have been made possible through the generosity of Gary M. Watts, MD and the Department of Radiology at Utah Valley Regional Medical Center.

We gratefully acknowledge the assistance of Nathan A. Jacobson, Scott R. Gunn, Sandra E. Sephton, Michelle Kidder, and Michael K. Visick for their help with dissections. Thanks go to Christopher H. Creek and Jessica Ridd for the art throughout the book. We also appreciate Focus Design and its employees for the layout and organization of this atlas. We are indebted to Douglas Morton and the personnel at Morton Publishing Company for the opportunity, encouragement, and support to prepare this atlas.

Photo Credits

Many of the photographs of living plants and animals were made possible because of the cooperation and generosity of the San Diego Zoo, San Diego Wild Animal Park, Sea World (San Diego, CA), Hogle Zoo (Salt Lake City, UT), and Aquatica (Orem, UT). We are especially appreciative to the professional biologists at these fine institutions.

We are appreciative of Dr. Wildford M. Hess and Dr. William B. Winborn for their help in obtaining photographs and photomicrographs. The electron micrographs are courtesy of Scott C. Miller. The microscope photo in Figure 1.2 was supplied by Leica Inc.

Figures 7.183, 7.186, 7.188, 7.189, 7.190, 8.4, 8.11, and 8.12 from *Comparative Anatomy: Manual of Vertebrate Dissection, 2nd Edition,* by Dale W. Fishbeck and Aurora Sebastiani. © 2008 Morton Publishing.

Figures 1.13, 4.20, 4.21, 4.22, 4.23, 4.24, 4.25, and 4.31 from *A Photographic Atlas for the Microbiology Laboratory, 3rd Edition,* by Michael J. Leboffe and Burton E. Pierce. © 2001 Morton Publishing.

Figures 8.110, 8.111, 8.112, 8.113, 8.114, 8.115, 8.116, 8.117, 8.118, 8.119, and 8.120 from *Mammalian Anatomy: The Cat, 2nd Edition,* by Aurora Sebastiani and Dale W. Fishbeck. ©2005 Morton Publishing.

Figures 8.63, 8.64, 8.65, 8.66, 8.67, 8.68, 8.69, 8.70, 8.71, and 8.72 from *A Dissection Guide and Atlas to the Rat,* by David G. Smith and Michael P. Schenk. © 2001 Morton Publishing.

Figures 8.75, 8.76, 8.77, 8.78 and 8.79 from *A Dissection Guide and Atlas to the Fetal Pig, 2nd Edition,* by David G. Smith and Michael P. Schenk. © 2003 Morton Publishing.

Book Team

Publisher: Doug Morton
Biology Editor: David Ferguson
Typography and Design: Focus Design
Cover: Bob Schram, Bookends Publication Design
Illustrations: Jessica Ridd

The fly agaric, or fly mushroom, *Amanita Muscaria,* is a common mushroom found throughout much of the Northern Hemisphere.

Table of Contents

All organisms are composed of one or more cells. *Cells* are the basic structural and functional units of organisms. A cell is a minute, membrane-enclosed, protoplasmic mass consisting of chromosomes surrounded by cytoplasm. Specific organelles are contained in the cytoplasm that function independently but in coordination with one another. Prokaryotic cells (fig. 1.1) and eukaryotic cells (figs. 1.3 and 1.18) are the two basic types.

Prokaryotic cells lack a membrane-bound nucleus, instead containing a single strand of *nucleic acid*. These cells contain few organelles. A rigid or semi-rigid cell wall provides shape to the cell outside the *cell (plasma) membrane*. Bacteria are examples of prokaryotic, single-celled organisms.

Eukaryotic cells contain a true *nucleus* with multiple chromosomes, have several types of specialized organelles, and have a differentially permeable cell membrane. Organisms consisting of eukaryotic cells include protozoa, fungi, algae, plants, and invertebrate and vertebrate animals.

Plant cells differ in some ways from other eukaryotic cells in that their cell walls contain *cellulose* for stiffness (fig. 1.3). Plant cells also contain vacuoles for water storage and membrane-bound *chloroplasts* with photosynthetic pigments for photosynthesis.

The *nucleus* is the large, spheroid body within the eukaryotic cell that contains the genetic material of the cell. The nucleus is enclosed by a double membrane called the *nuclear membrane*, or *nuclear envelope*. The *nucleolus* is a dense, nonmembranous body composed of protein and RNA molecules. The chromatin are fibers of protein and DNA molecules that make up a eukaryotic chromosome. Prior to cellular division, the chromatin shortens and coils into rod-shaped *chromosomes*. Chromosomes consist of DNA and structural proteins called *histones*.

The *cytoplasm* of the eukaryotic cell is the medium between the nuclear membrane and the cell membrane. *Organelles* are small membrane-bound structures within the cytoplasm. The cellular functions carried out by organelles are referred to as *metabolism*. The structure and function of the nucleus and principal organelles are listed in table 1.1. In order for cells to remain alive, metabolize, and maintain homeostasis, they must have access to nutrients and respiratory gases, be able to eliminate wastes, and be in a constant, protective environment.

The *cell membrane* is composed of phospholipid, protein, and carbohydrate molecules. The cell membrane gives form to a cell and controls the passage of material into and out of a cell. More specifically, the proteins in the cell membrane provide:
1. structural support;
2. a mechanism of molecule transport across the membrane;
3. enzymatic control of chemical reactions;
4. receptors for hormones and other regulatory molecules; and

5. cellular markers (antigens), which identify the blood and tissue type.

The carbohydrate molecules:
1. repel negative objects due to their negative charge;
2. act as receptors for hormones and other regulatory molecules;
3. form specific cell markers that enable like cells to attach and aggregate into tissues; and
4. enter into immune reactions.

Tissues are groups of similar cells that perform specific functions (see fig. 1.9). A flowering plant, for example, is composed of three tissue systems:
1. the *ground tissue system*, providing support, regeneration, respiration, photosynthesis, and storage;
2. the *vascular tissue system*, providing conduction passageways through the plant; and
3. the *dermal tissue system*, providing protection to the plant.

The tissues of the body of a multicellular animal are classified into four principal types (see fig. 1.36):
1. *epithelial tissue* covers body and organ surfaces, lines body cavities and lumina (hollow portions of body tubes), and forms various glands;
2. *connective tissue* binds, supports, and protects body parts;
3. *muscle tissue* contracts to produce movements; and
4. *nervous tissue* initiates and transmits nerve impulses.

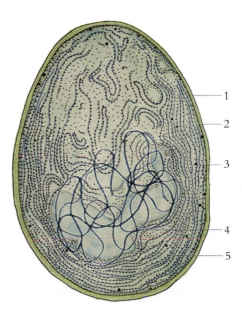

Figure 1.1 A prokaryotic cell.
1. Cell wall
2. Cell (plasma) membrane
3. Ribosomes
4. Circular molecule of DNA
5. Thylakoid membranes

Table 1.1 Structure and Function of Eukaryotic Cellular Components

Component	Structure	Function
Cell (plasma) membrane	Composed of protein and phospholipid molecules	Provides form to cell; controls passage of materials into and out of cell
Cell Wall	Cellulose fibrils	Provides structure and rigidity to plant cell
Cytoplasm	Fluid to jelly-like substance	Serves as suspending medium for organelles and dissolved molecules
Endoplasmic reticulum	Interconnecting membrane-lined channels	Enables cell transport and processing of metabolic chemicals
Ribosome	Granules of nucleic acid (RNA) and protein	Synthesizes protein
Mitochondrion	Double-membraned sac with cristae (chambers)	Assembles ATP (cellular respiration)
Golgi complex	Flattened membrane-lined chambers	Synthesize carbohydrates and packages molecules for secretion
Lysosome	Membrane-surrounded sac of enzymes	Digests foreign molecules and worn cells
Centrosome	Mass of protein that may contain rod-like centrioles	Organizes spindle fibers and assists mitosis and meiosis
Vacuole	Membranous sac	Stores and excretes substances within the cytoplasm, regulates cellular turgor pressure
Microfibril and microtubule	Protein strands and tubes	Forms cytoskeleton, supports cytoplasm and transports materials
Cilium and flagellum	Cytoplasmic extensions from cell; containing microtubules	Movements of particles along cell surface, or cell movement
Nucleus	Nuclear envelope (membrane), nucleolus, and chromatin (DNA)	Contains genetic code that directs cell activity; forms ribosomes
Chloroplast	Inner (grana) membrane within outer membrane	Involved in photosynthesis

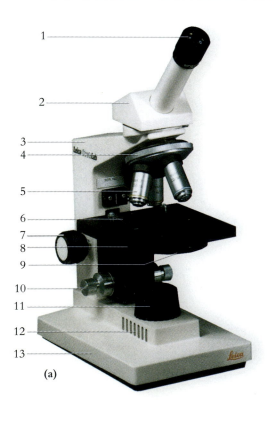

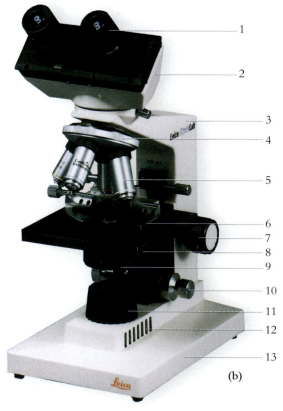

Figure 1.2 (a) A compound monocular microscope, and (b) compound binocular microscope.

1. Eyepiece (ocular)
2. Body
3. Arm
4. Nosepiece
5. Objective
6. Stage clip
7. Focus adjustment knob
8. Stage
9. Condenser
10. Fine focus adjustment knob
11. Collector lens with field diaphragm
12. Illuminator (inside)
13. Base

(a)

(b)

Plant cell and tissues

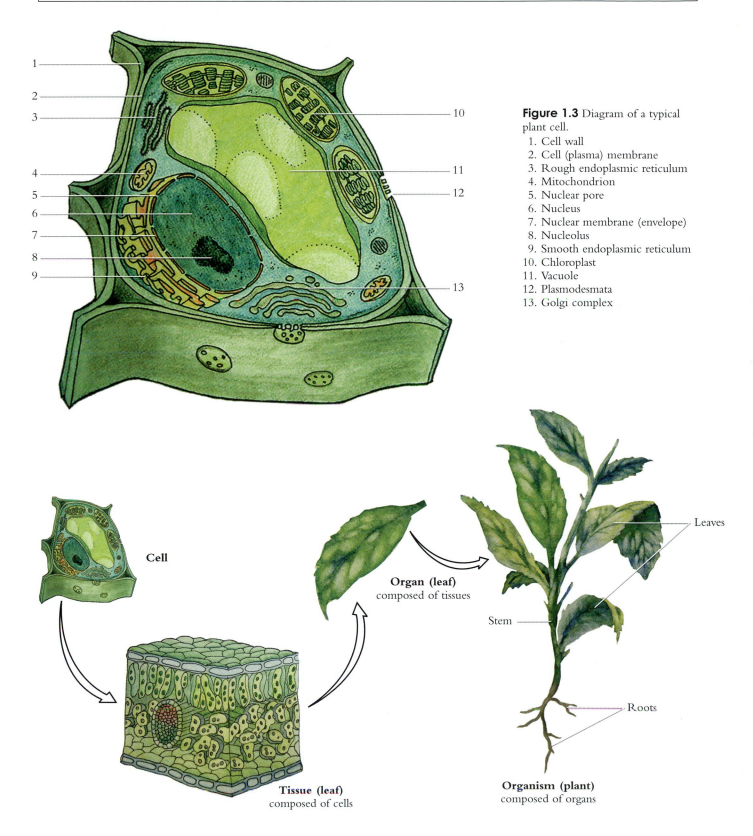

Figure 1.3 Diagram of a typical plant cell.
1. Cell wall
2. Cell (plasma) membrane
3. Rough endoplasmic reticulum
4. Mitochondrion
5. Nuclear pore
6. Nucleus
7. Nuclear membrane (envelope)
8. Nucleolus
9. Smooth endoplasmic reticulum
10. Chloroplast
11. Vacuole
12. Plasmodesmata
13. Golgi complex

Cell

Tissue (leaf)
composed of cells

Organ (leaf)
composed of tissues

Organism (plant)
composed of organs

Leaves

Stem

Roots

Figure 1.4 The structural levels of plant organization.

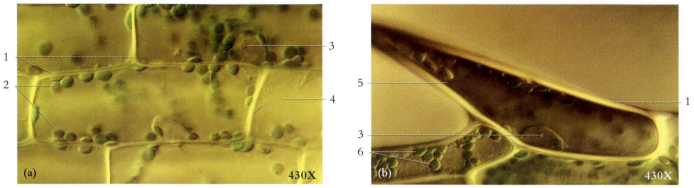

Figure 1.5 A live *Elodea* leaf: cells (a) photographed at the center of the leaf and (b) at the edge of the leaf.
1. Cell wall   3. Nucleus   5. Spine-shaped cell on   6. Chloroplasts
2. Chloroplasts   4. Vacuole   exposed edge of leaf

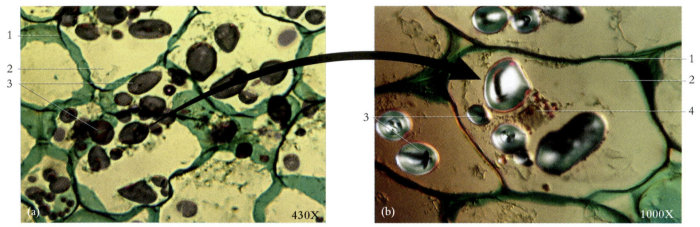

Figure 1.6 (a) Cells of a potato, *Solanum tuberosum*, showing starch grains at a low magnification, and (b) at a high magnification. Food is stored as starch in potato cells, which is deposited in organelles called amyloplasts.
1. Cell wall   2. Cytoplasm   3. Starch grains   4. Nucleus

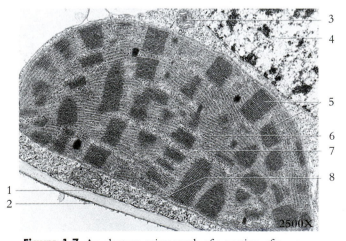

Figure 1.7 An electron micrograph of a portion of a sugar cane leaf cell.
1. Cell membrane   5. Grana
2. Cell wall   6. Stroma
3. Mitochondrion   7. Thylakoid membrane
4. Nucleus   8. Chloroplast envelope

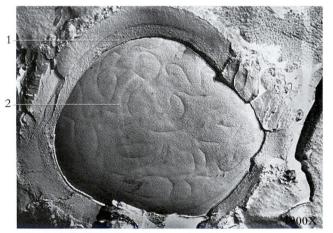

Figure 1.8 A fractured barley smut spore.
1. Cell wall   2. Cell membrane

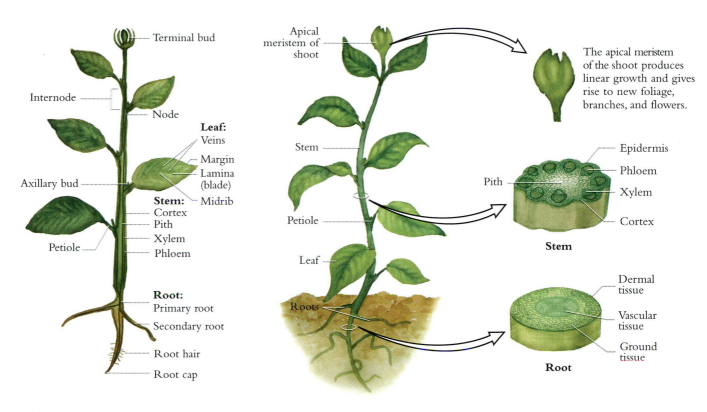

Figure 1.9 A diagram illustrating the anatomy and the principal organs and tissues of a typical dicot.

The following labels appear in the diagram:

Terminal bud
Internode
Node
Leaf:
Veins
Margin
Lamina (blade)
Stem: Midrib
Cortex
Pith
Xylem
Phloem
Axillary bud
Petiole
Root:
Primary root
Secondary root
Root hair
Root cap

Apical meristem of shoot
The apical meristem of the shoot produces linear growth and gives rise to new foliage, branches, and flowers.
Stem
Petiole
Leaf
Roots

Epidermis
Phloem
Xylem
Cortex
Pith
Stem

Dermal tissue
Vascular tissue
Ground tissue
Root

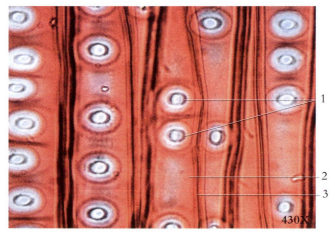

Figure 1.10 A longitudinal section through the xylem of a pine, *Pinus*, showing tracheid cells with prominent bordered pits.

1. Bordered pits
2. Tracheid cell
3. Cell wall

430X

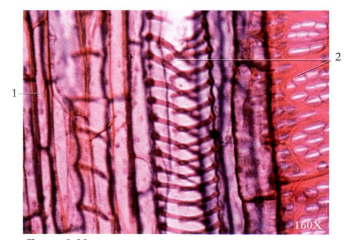

Figure 1.11 A longitudinal section through the xylem of a squash stem, *Cucurbita maxima*. The vessel elements shown here have several different patterns of wall thickenings.

1. Parenchyma
2. Vessel elements

160X

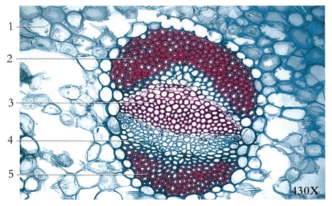

Figure 1.12 A section through a leaf of the venus flytrap, *Dionaea muscipula*, showing epidermal cells with a digestive gland. The gland is composed of secretory parenchyma cells.
1. Epidermis   2. Gland

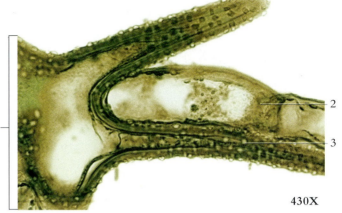

Figure 1.13 An astrosclereid in the petiole of a pondlily, *Nuphar*.
1. Astrosclereid   3. Crystals in cell wall
2. Parenchyma cell

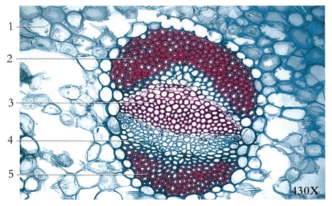

Figure 1.14 A transverse section through the leaf of a yucca, *Yucca brevifolia*, showing a vascular bundle (vein). Note the prominent sclerenchyma tissue forming caps on both sides of the bundle.
1. Leaf parenchyma   3. Xylem
2. Leaf sclerenchyma   4. Phloem
   (bundle cap)   5. Bundle cap

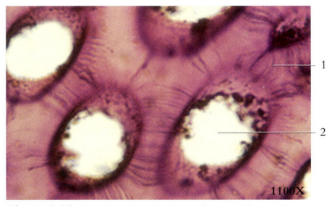

Figure 1.15 A section through the endosperm tissue of a persimmon, *Diospyros virginiana*. These thick-walled cells are actually parenchyma cells. Cytoplasmic connections, or plasmodesmata, are evident between cells.
1. Plasmodesmata   2. Cell lumen (interior space)

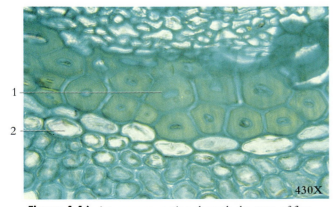

Figure 1.16 A transverse section through the stem of flax, *Linum*. Note the thick-walled fibers as compared to the thin-walled parenchyma cells.
1. Fibers   2. Parenchyma cell

Figure 1.17 A section through the stem of a wax plant, *Hoya carnosa*. Thick-walled sclereids (stone cells) are evident.
1. Parenchyma cell   2. Sclereid (stone cell)
   containing starch grains

Animal cell and tissues

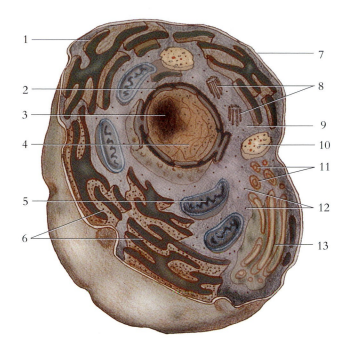

Figure 1.18 A typical animal cell.

1. Smooth endoplasmic
   reticulum
2. Nuclear membrane
3. Nucleolus
4. Nucleoplasm
5. Mitochondrion
6. Rough endoplasmic
   reticulum

7. Cell membrane
8. Centrosomes
9. Cytoplasm
10. Lysosome
11. Vesicles
12. Ribosomes
13. Golgi complex

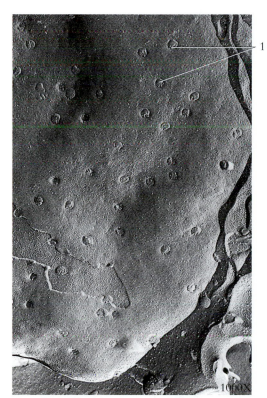

Figure 1.19 An electron micrograph of a
freeze-fractured nuclear envelope showing the
nuclear pores.

1. Nuclear pores

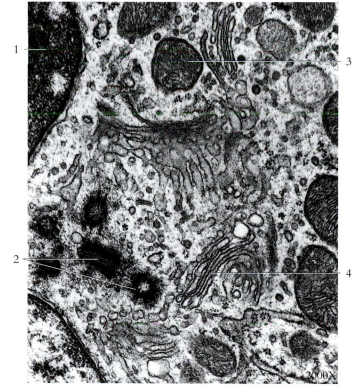

Figure 1.20 An electron micrograph of various organelles.

1. Nucleus
2. Centrioles

3. Mitochondrion
4. Golgi complex

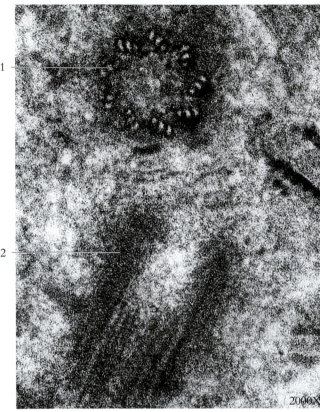

2000X

Figure 1.21 An electron micrograph of centrioles. The centrioles are positioned at right angles to one another.

1. Centriole (shown in transverse section)
2. Centriole (shown in longitudinal section)

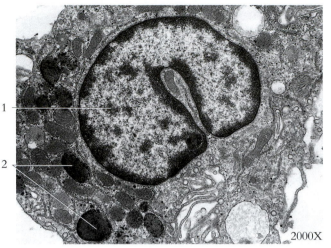

2000X

Figure 1.22 An electron micrograph of lysosomes.

1. Nucleus   2. Lysosomes

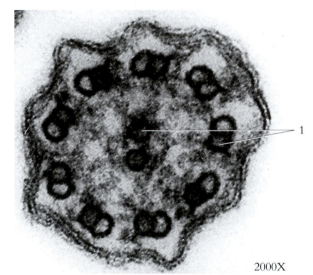

2000X

Figure 1.24 An electron micrograph of cilia (transverse section) showing the characteristic "9 + 2" arrangement of microtubules in the transverse sections.

1. Microtubules

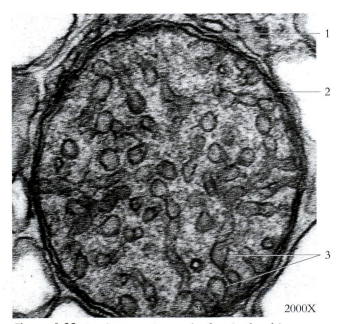

2000X

Figure 1.23 An electron micrograph of a mitochondrion.

1. Outer membrane   3. Inner membranes
2. Crista

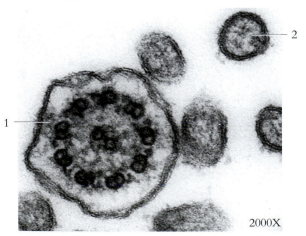

2000X

Figure 1.25 An electron micrograph showing the difference between a microvillus and a cilium.

1. Cilium   2. Microvillus

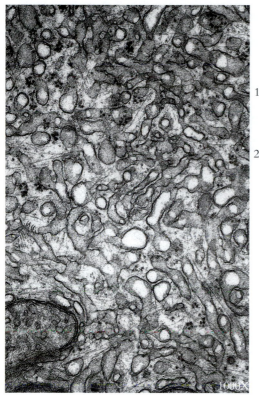

Figure 1.26 An electron micrograph of smooth endoplasmic reticulum from the testis.

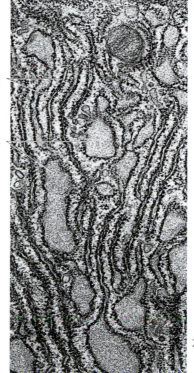

Figure 1.27 An electron micrograph of rough endoplasmic reticulum.

1. Ribosomes
2. Cisternae

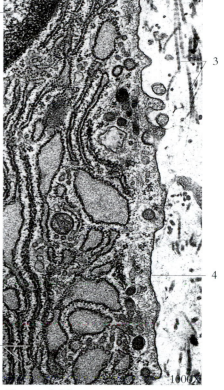

Figure 1.28 An electron micrograph of rough endoplasmic reticulum secreting collagenous filaments to the outside of the cell.

1. Nucleus
2. Rough endoplasmic reticulum
3. Collagenous filaments
4. Cell membrane

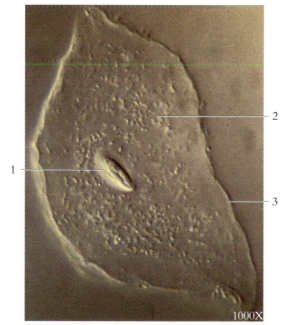

Figure 1.29 An epithelial cell from a cheek scraping.

1. Nucleus
2. Cytoplasm
3. Cell membrane

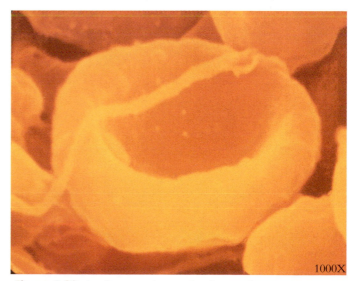

Figure 1.30 An electron micrograph of an erythrocyte (red blood cell).

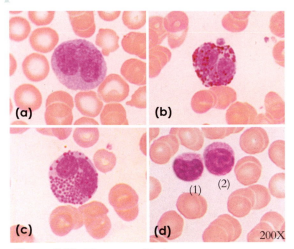

Figure 1.31 The types of leukocytes.

  (a) Neutrophil   (d) Lymphocyte (1)
  (b) Basophil   Monocyte (2)
  (c) Eosinophil

Figure 1.32 An electron micrograph of a capillary containing an erythrocyte.

  1. Lumen of capillary   3. Endothelial cell
  2. Nucleus of endothelial cell   4. Erythrocyte

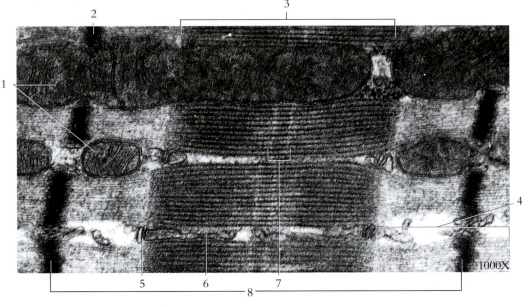

Figure 1.33 An electron micrograph of a skeletal muscle myofibril, showing the striations.

  1. Mitochondria
  2. Z line
  3. A band
  4. I band
  5. T-tubule
  6. Sarcoplasmic reticulum
  7. H band
  8. Sarcomere

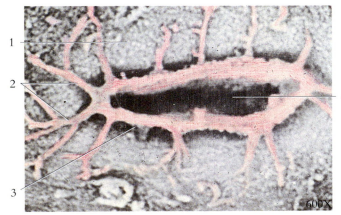

Figure 1.34 An electron micrograph of an osteocyte (bone cell) in cortical bone matrix.

  1. Bone matrix   3. Lacuna
  2. Canaliculi   4. Osteocyte

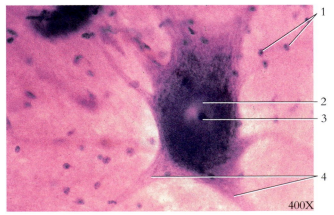

Figure 1.35 A neuron smear.

  1. Nuclei of surrounding neuroglial cells
  2. Nucleus of neuron
  3. Nucleolus of neuron
  4. Dendrites of neuron

Epithelial Tissue

Epithelial tissue covers the outside of
the body and lines all organs. Its primary
function is to provide protection.

— Single cell layer

Simple squamous epithelium

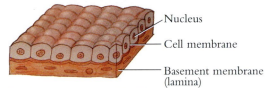

— Nucleus

— Cell membrane

— Basement membrane
(lamina)

Simple cuboidal epithelium

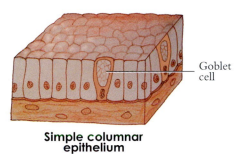

— Goblet
cell

**Simple columnar
epithelium**

Connective Tissue

Connective tissue functions as a binding
and supportive tissue for all other tissues
in the organism.

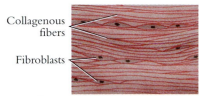

Collagenous
fibers —

Fibroblasts —

Dense regular connective tissue

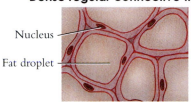

Nucleus —

Fat droplet —

Adipose tissue

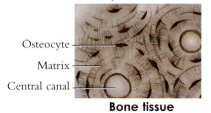

Osteocyte —

Matrix —

Central canal —

Bone tissue

Muscle Tissue

Muscle tissue is a tissue adapted to contract.
Muscles provide movement and functionality
to the organism.

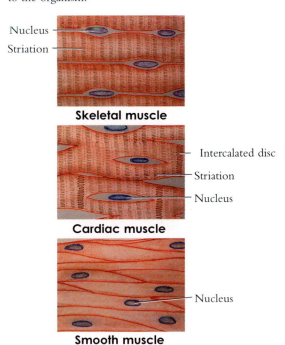

Nucleus —

Striation —

Skeletal muscle

— Intercalated disc

— Striation

— Nucleus

Cardiac muscle

— Nucleus

Smooth muscle

Nervous Tissue

Nervous tissue functions to receive stimuli and transmits signals from one part of
the organism to another.

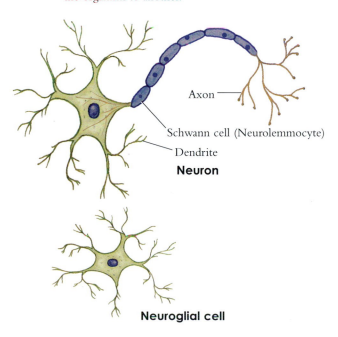

— Axon

Schwann cell (Neurolemmocyte)

Dendrite

Neuron

Neuroglial cell

Figure 1.36 Some examples of animal tissues.

Figure 1.37 Simple squamous epithelium.
1. Single layer of flattened cells

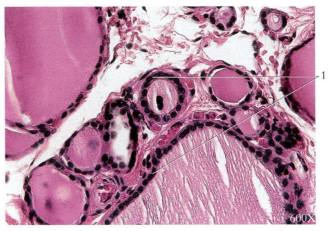

Figure 1.38 Simple cuboidal epithelium.
1. Single layer of cells with round nuclei

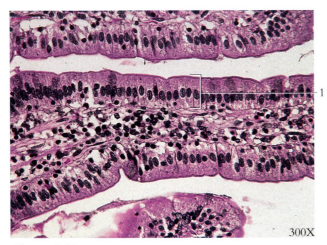

Figure 1.39 Simple columnar epithelium.
1. Single layer of cells with oval nuclei

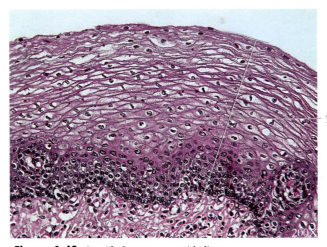

Figure 1.40 Stratified squamous epithelium.
1. Multiple layers of cells, which are flattened at the upper layer

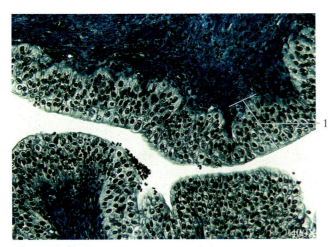

Figure 1.41 Stratified columnar epithelium.
1. Cells are balloon-like at surface

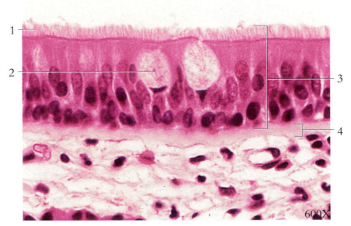

Figure 1.42 Pseudostratified columnar epithelium.
1. Cilia
2. Goblet cell
3. Pseudostratified columnar epithelium
4. Basement membrane

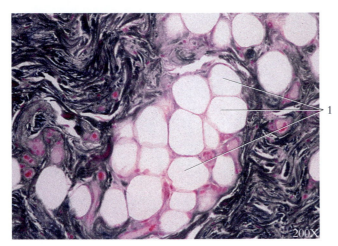

Figure 1.43 Adipose connective tissue.
1. Adipocytes (adipose cells)

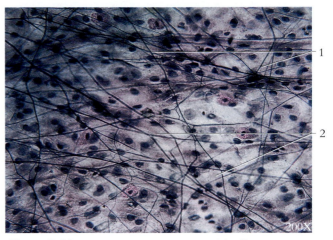

Figure 1.44 Loose connective tissue stained for fibers.
1. Elastic fibers (black)
2. Collagen fibers (pink)

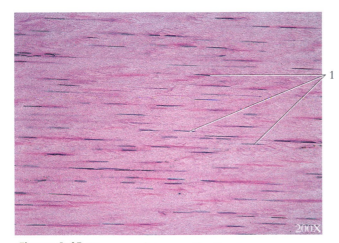

Figure 1.45 Dense regular connective tissue.
1. Nuclei of fibroblasts arranged in parallel rows

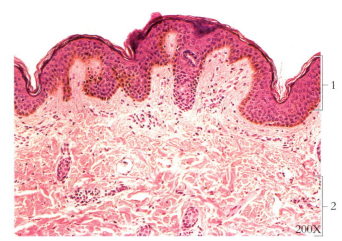

Figure 1.46 Dense irregular connective tissue.
1. Epidermis
2. Dense irregular connective tissue (reticular layer of dermis)

Figure 1.47 An electron micrograph of dense irregular connective tissue.
1. Collagenous fibers

Figure 1.48 Reticular connective tissue.
1. Reticular fibers

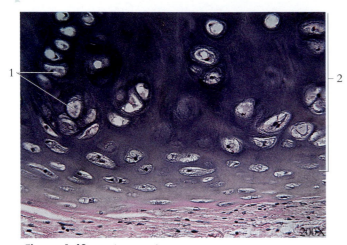

Figure 1.49 Hyaline cartilage.
1. Chondrocytes
2. Hyaline cartilage

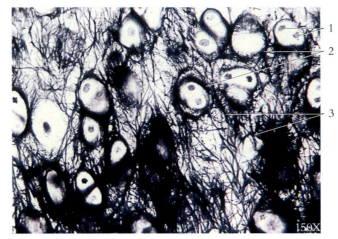

Figure 1.50 Elastic cartilage.
1. Chondrocytes   3. Elastic fibers
2. Lacunae

Figure 1.51 Fibrocartilage.
1. Chondrocytes arranged in a row

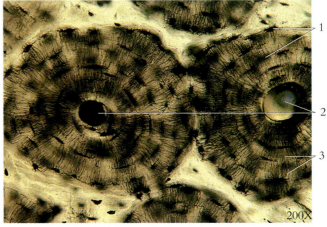

Figure 1.52 A transverse section of two osteons.
1. Lacunae   3. Lamellae
2. Central (Haversian) canals

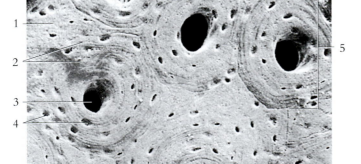

Figure 1.53 An electron micrograph of bone tissue.
1. Interstitial lamellae   4. Lacunae
2. Lamellae   5. Osteon (Haversian
3. Central canal   system)
   (Haversian canal)

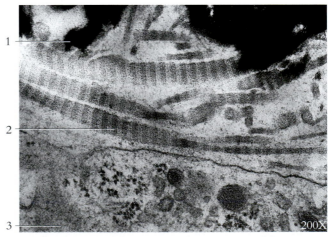

Figure 1.54 An electron micrograph of bone tissue formation.
1. Bone mineral (calcium salts stain black)
2. Collagenous filament (distinct banding pattern)
3. Collagen secreting osteoblasts

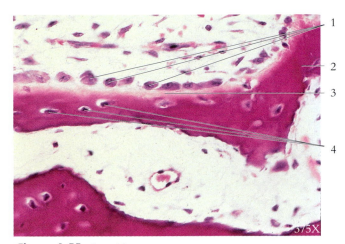

Figure 1.55 Osteoblasts.
1. Osteoblasts   3. Osteoid
2. Bone   4. Osteocytes

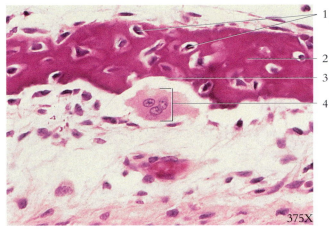

Figure 1.56 Osteoclast.
1. Osteocytes   4. Osteoclast in
2. Bone   Howship's lacuna
3. Howship's lacuna

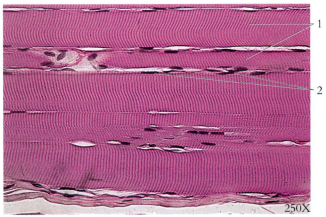

Figure 1.57 A longitudinal section of skeletal muscle tissue.
1. Skeletal muscle cells (note striations)
2. Multiple nuclei in periphery of cell

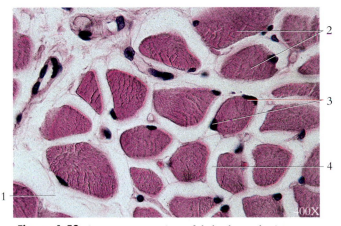

Figure 1.58 A transverse section of skeletal muscle tissue.
1. Perimysium (surrounds bundles of cells)
2. Skeletal muscle cells
3. Nuclei in periphery of cell
4. Endomysium (surrounds cells)

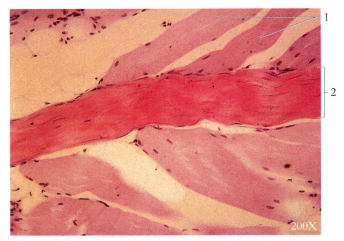

Figure 1.59 The attachment of skeletal muscle to tendon.
1. Skeletal muscle
2. Dense regular connective tissue (tendon)

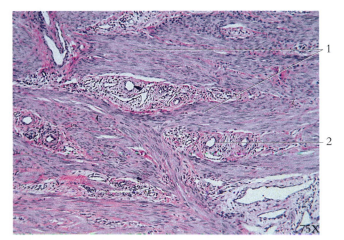

Figure 1.60 Smooth muscle tissue.
1. Smooth muscle
2. Blood vessel

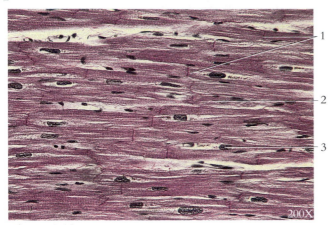

Figure 1.61 Cardiac muscle tissue.
1. Intercalated discs
2. Light-staining perinuclear sarcoplasm
3. Nucleus in center of cell

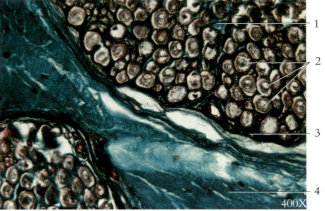

Figure 1.62 A transverse section of a nerve.
1. Endoneurium   3. Perineurium
2. Axons   4. Epineurium

Figure 1.63 A longitudinal section of axons.
1. Myelin sheath
2. Neurofibril nodes (nodes of Ranvier)

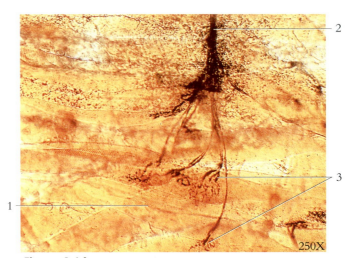

Figure 1.64 Neuromuscular junction.
1. Skeletal muscle   2. Motor nerve
   fiber   3. Motor end plates

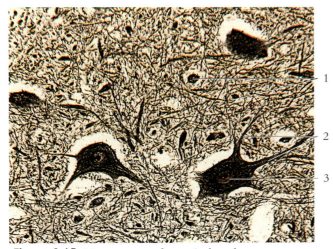

Figure 1.65 Motor neurons from spinal cord.
1. Neuroglia cells
2. Dendrites
3. Nucleus

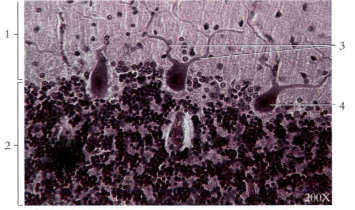

Figure 1.66 Purkinje neurons from the cerebellum.
1. Molecular layer of cerebellar cortex
2. Granular layer of cerebellar cortex
3. Dendrites of Purkinje cell
4. Purkinje cell body

The term *cell cycle* refers to how a multicellular organism develops, grows, and maintains and repairs body tissues. In the cell cycle, each new cell receives a complete copy of all genetic information in the parent cell, and the cytoplasmic substances and organelles to carry out hereditary instructions.

The animal cell cycle (see fig. 2.3 and 2.7) is divided into: 1) interphase, which includes G1, S, and G2 phases; and 2) mitosis, which includes prophase, metaphase, anaphase, and telophase. *Interphase* is the interval between successive cell divisions during which the cell is metabolizing and the chromosomes are directing RNA synthesis. The *G1 phase* is the first growth phase, the *S phase* is when DNA is replicated, and the *G2 phase* is the second growth phase. *Mitosis* (also known as karyokinesis) is the division of the nuclear parts of a cell to form two daughter nuclei with the same number of chromosomes as the original nucleus.

Like the animal cell cycle, the plant cell cycle consists of growth, synthesis, mitosis, and cytokinesis. *Growth* is the increase in cellular mass as the result of metabolism; *synthesis* is the production of DNA and RNA to regulate cellular activity; mitosis is the splitting of the nucleus and the equal separation of the chromatids; and cytokinesis is the division of the cytoplasm that accompanies mitosis.

Unlike animal cells, plant cells have a rigid cell wall that does not cleave during cytokinesis. Instead, a new cell wall is constructed between the daughter cells (see fig. 2.5 and 2.6). Furthermore, many land plants do not have centrioles for the attachment of spindles. The microtubules in these plants form a barrel-shaped anastral spindle at each pole. Mitosis and cytokinesis in plants occurs in basically the same sequence as these processes in animal cells.

Asexual reproduction is propagation without sex; that is, the production of new individuals by processes that do not involve *gametes* (sex cells). Asexual reproduction occurs in a variety of microorganisms, fungi, plants, and animals, wherein a single parent produces offspring with characteristics identical to itself. Asexual reproduction is not dependent on the presence of other individuals. No egg or sperm is required. In asexual reproduction, all the offspring are genetically identical (except for mutants). Types of asexual reproduction include:

1. *fission*—a single cell divides to form two separate cells (bacteria, protozoans, and other one-celled organisms);
2. *sporulation*—multiple fission; many cells are formed and join together in a cystlike structure (protozoans and fungi);
3. *budding*—buds develop into organisms like the parent and then detach themselves (hydras, yeast, certain plants); and
4. *fragmentation*—organisms break into two or more parts, and each part is capable of becoming a complete organism (algae, flatworms, echinoderms).

Sexual reproduction is propagation of new organisms through the union of genetic material from two parents. Sexual reproduction usually involves the fusion of haploid gametes (such as sperm and egg cells) during fertilization to form a zygote.

The major biological difference between sexual and asexual reproduction is that sexual reproduction produces genetic variation in the offspring. The combining of genetic material from the gametes produces offspring that are different from either parent and contain new combinations of characteristics. This may increase the ability of the species to survive environmental changes or to reproduce in new habitats. The only genetic variation that can arise in asexual reproduction comes from mutations.

Figure 2.1 Sexual reproduction. A pair of California quail, *Callipepla californica*, in early spring.

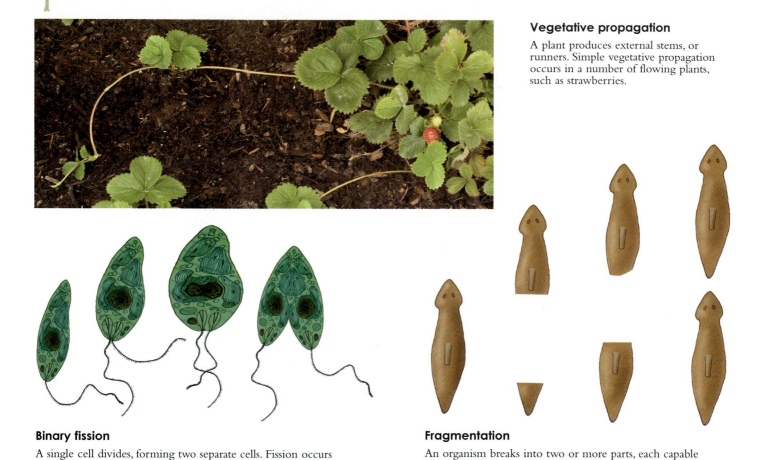

Vegetative propagation

A plant produces external stems, or runners. Simple vegetative propagation occurs in a number of flowing plants, such as strawberries.

Binary fission

A single cell divides, forming two separate cells. Fission occurs in bacteria, protozoans, and other single-celled organisms.

Fragmentation

An organism breaks into two or more parts, each capable of becoming a complete organism. Fragmentation occurs in flatworms and echinoderms.

Figure 2.2 Types of asexual reproduction: vegetative propagation, binary fission, and fragmentation.

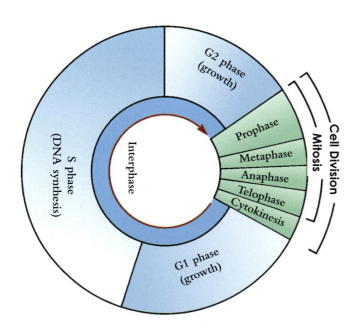

Figure 2.3 The animal cell cycle.

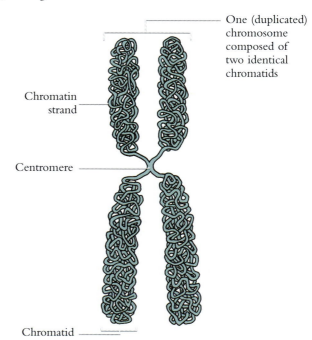

Figure 2.4 Each duplicated chromosome consists of two identical chromatids attached at the centrally located and constricted centromere.

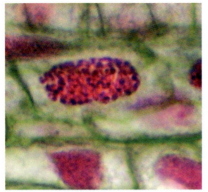

Early prophase — Chromatin begins to condense to form chromosomes.

Late prophase — Nuclear envelope is intact, and chromatin condenses into chromosomes.

Early metaphase — Duplicated chromosomes are each made up of two chromatids, at equatorial plane.

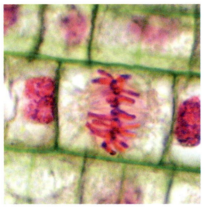

Late metaphase — Duplicated chromosomes are each made up of two chromatids, at equatorial plane.

Early anaphase — Sister chromatids are beginning to separate into daughter chromosomes.

Late anaphase — Daughter chromosomes are nearing poles.

1

Telophase — Daughter chromosomes are at poles, and cell plate is forming.
    1. Cell plate

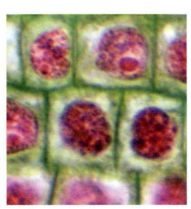

Interphase — Two daughter cells result from cytokinesis.

Figure 2.5 The stages of mitosis in Hyacinth, *Hyacinthus*, root tip. (all 430X)

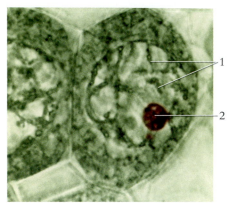

Prophase I — Each chromosome consists of two chromatids joined by a centromere.
1. Chromatids
2. Nucleolus

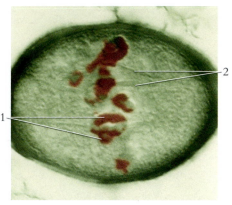

Metaphase I — Chromosome pairs align at the equator.
1. Chromosome pairs at equator
2. Spindle fibers

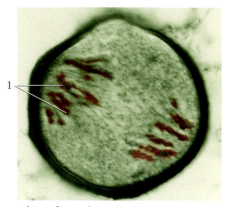

Anaphase I — No division at the centromeres occurs as the chromosomes separate, so one entire chromosome goes to each pole.
1. Chromosomes (two chromatids each)

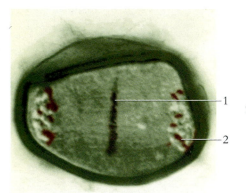

Telophase I — Chromosomes lengthen and become less distinct. The cell plate (in some plants) forms between forming cells.
1. Cell plate (new cell wall)
2. Chromosome

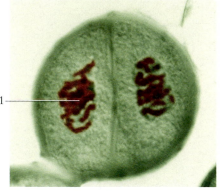

Prophase II — Chromosomes condense as in prophase I.
1. Chromosomes

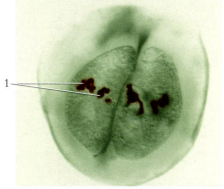

Metaphase II — Chromosomes align on the equator, and spindle fibers attach to the centromeres. This is similar to metaphase in mitosis.
1. Chromosomes

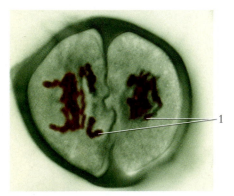

Anaphase II — Chromatids separate and each is pulled to an opposite pole.
1. Chromatids

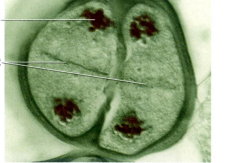

Telophase II — Cell division is complete and cell walls of four haploid cells are formed.
1. Chromatids
2. New cell walls (cell plates)

Figure 2.6 The stages of meiosis in lily microsporocytes to form microspores. 1000X

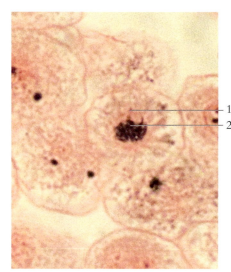

Prophase

Each chromosome consists of two chromatids joined by a centromere. Spindle fibers extend from each centriole.

1. Aster around centriole
2. Chromosomes

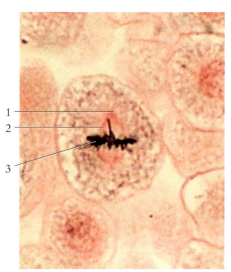

Metaphase

The chromosomes are positioned at the equator. The spindle fibers from each centriole attach to the centromeres.

1. Aster around centriole
2. Spindle fibers
3. Chromosomes at equator

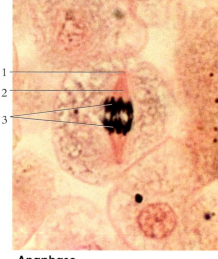

Anaphase

The centromeres split, and the sister chromatids separate as each is pulled to an opposite pole.

1. Aster around centriole
2. Spindle fibers
3. Separating chromosomes

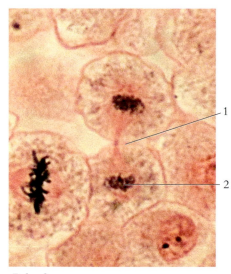

Telophase

The chromosomes lengthen and become less distinct. The cell membrane forms between the forming daughter cells.

1. New cell membrane
2. Newly forming nucleus

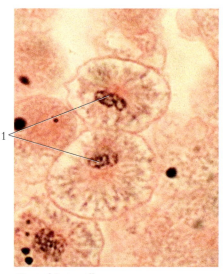

Daughter cells

The single chromosomes (former chromatids—see anaphase) continue to lengthen as the nuclear membrane reforms. Cell division is complete and the newly formed cells grow and mature.

1. Daughter nuclei

Figure 2.7 The stages of animal cell mitosis followed by cytokinesis. Whitefish blastula. 500X

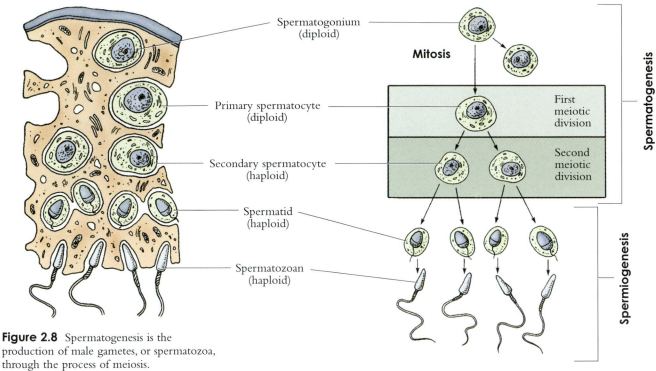

Spermatogonium
(diploid)

Mitosis

Primary spermatocyte
(diploid)

First meiotic division

Secondary spermatocyte
(haploid)

Second meiotic division

Spermatid
(haploid)

Spermatozoan
(haploid)

Spermatogenesis

Spermiogenesis

Figure 2.8 Spermatogenesis is the production of male gametes, or spermatozoa, through the process of meiosis.

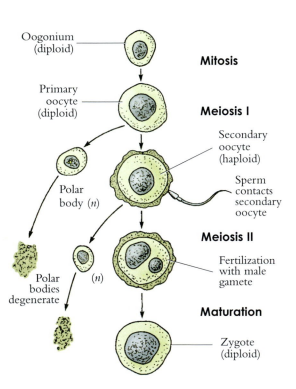

Oogonium
(diploid)

Mitosis

Primary oocyte
(diploid)

Meiosis I

Secondary oocyte
(haploid)

Polar body (n)

Sperm contacts secondary oocyte

Polar bodies degenerate

(n)

Meiosis II

Fertilization with male gamete

Maturation

Zygote
(diploid)

Figure 2.9 Oogenesis is the production of female gametes, or ova, through the process of meiosis.

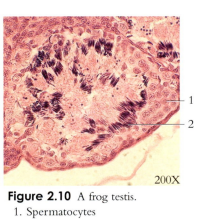

Figure 2.10 A frog testis.
1. Spermatocytes
2. Developing sperm

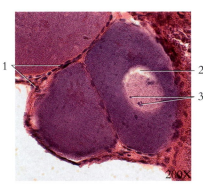

Figure 2.11 A frog ovary.
1. Follicle cells
2. Germinal vesicle
3. Nucleoli

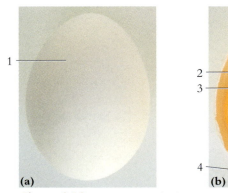

(a)

(b)

Figure 2.12 (a) An intact chicken egg and (b) a portion of the shell removed, exposing the internal structures.

1. Shell
2. Vitelline membrane
3. Yolk
4. Shell membrane

5. Shell
6. Albumen (egg white)
7. Chalaza (dense albumen)
8. Air space

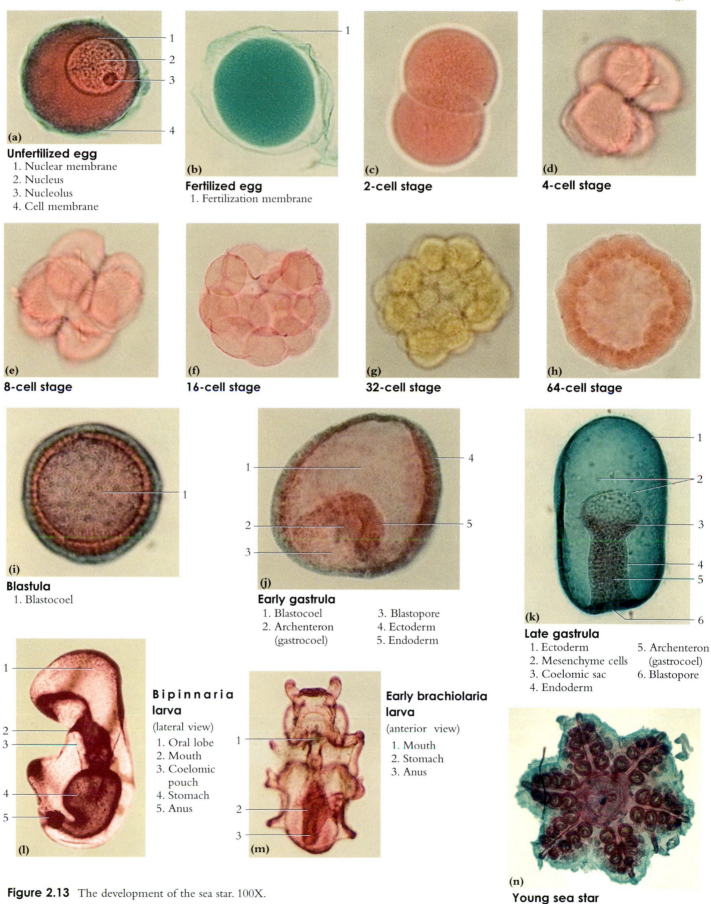

Unfertilized egg
1. Nuclear membrane
2. Nucleus
3. Nucleolus
4. Cell membrane

(a)

(b)

Fertilized egg
1. Fertilization membrane

(c)

2-cell stage

(d)

4-cell stage

(e)

8-cell stage

(f)

16-cell stage

(g)

32-cell stage

(h)

64-cell stage

(i)

Blastula
1. Blastocoel

(j)

Early gastrula
1. Blastocoel   3. Blastopore
2. Archenteron   4. Ectoderm
   (gastrocoel)   5. Endoderm

(k)

Late gastrula
1. Ectoderm   5. Archenteron
2. Mesenchyme cells   (gastrocoel)
3. Coelomic sac   6. Blastopore
4. Endoderm

(l)

Bipinnaria larva
(lateral view)
1. Oral lobe
2. Mouth
3. Coelomic pouch
4. Stomach
5. Anus

(m)

Early brachiolaria larva
(anterior view)
1. Mouth
2. Stomach
3. Anus

(n)

Young sea star

Figure 2.13 The development of the sea star. 100X.

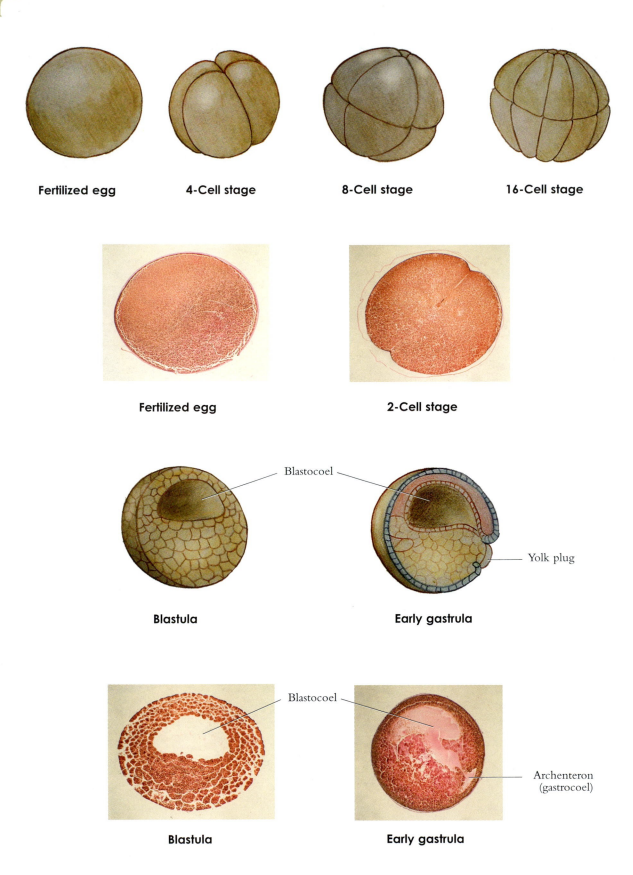

Fertilized egg   **4-Cell stage**   **8-Cell stage**   **16-Cell stage**

Fertilized egg   **2-Cell stage**

Blastocoel

Blastocoel

Yolk plug

Blastula   **Early gastrula**

Blastocoel   Blastocoel

Archenteron
(gastrocoel)

Blastula   **Early gastrula**

Figure 2.14 Frog development from fertilized egg to early gastrula, shown in diagram and photomicrographs. 100X.

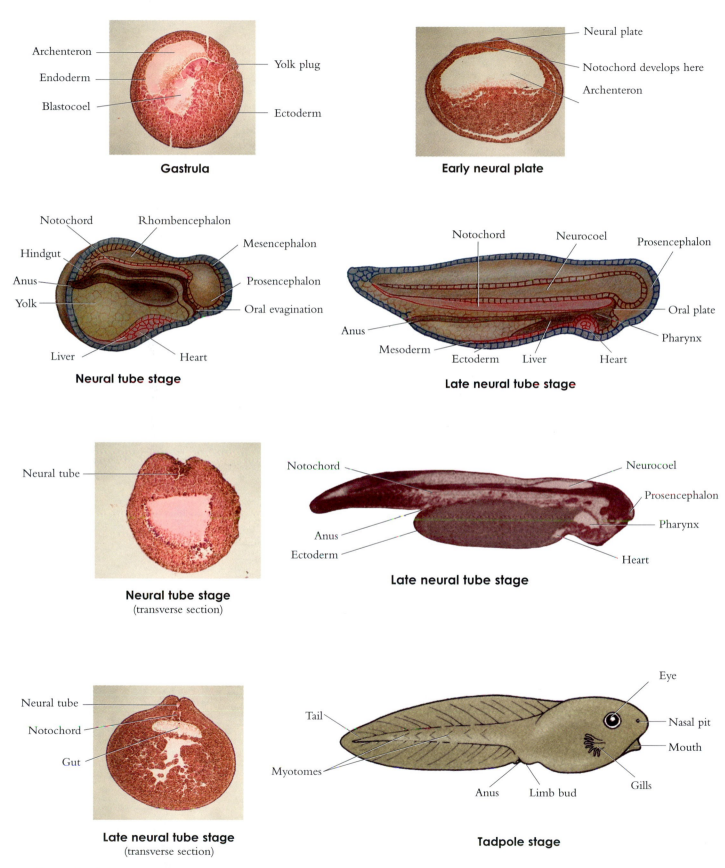

Archenteron
Endoderm
Blastocoel
Yolk plug
Ectoderm
Gastrula

Neural plate
Notochord develops here
Archenteron
Early neural plate

Notochord
Rhombencephalon
Hindgut
Mesencephalon
Anus
Prosencephalon
Yolk
Oral evagination
Liver
Heart
Neural tube stage

Notochord
Neurocoel
Prosencephalon
Anus
Oral plate
Mesoderm
Pharynx
Ectoderm
Liver
Heart
Late neural tube stage

Neural tube
Neural tube stage
(transverse section)

Notochord
Neurocoel
Prosencephalon
Anus
Pharynx
Ectoderm
Heart
Late neural tube stage

Neural tube
Notochord
Gut
Late neural tube stage
(transverse section)

Eye
Tail
Nasal pit
Mouth
Myotomes
Gills
Anus
Limb bud
Tadpole stage

Figure 2.15 Frog development from gastrula to tadpole, shown in diagram and photomicrographs. 100X.

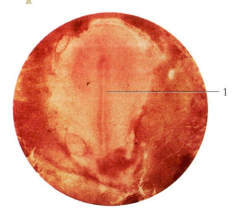

13-hour stage

1. Embryo main body
   formation

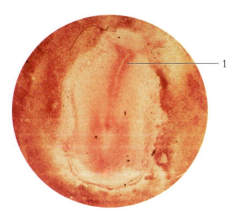

18-hour stage

1. Neurulation
   beginning

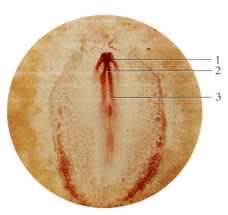

21-hour stage

1. Head fold
2. Neural fold
3. Muscle plate (somites)

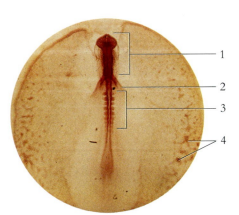

28-hour stage

1. Head fold and brain
2. Artery formation
3. Muscle plate (somites)
4. Blood vessel formation

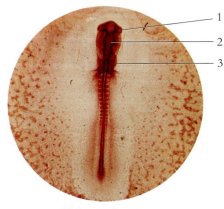

38-hour stage

1. Optic vesicle
2. Brain with 5 regions
3. Heart

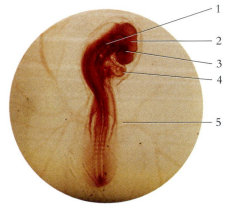

48-hour stage

1. Ear
2. Brain
3. Eye
4. Heart
5. Artery

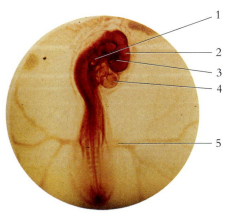

56-hour stage

1. Ear
2. Brain
3. Eye
4. Heart
5. Artery

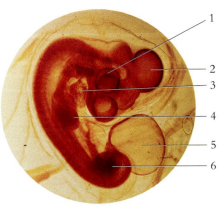

96-hour stage

1. Eye
2. Mesencephalon
3. Heart
4. Wing formation
5. Fecal sac (allantois)
6. Leg formation

Figure 2.16 The stages of
chick development. 20X

Bacteria range between 1 and 50 μm in width or diameter. The morphological appearance of bacteria may be spiral (spirillum), spherical (coccus), or rod-shaped (bacillus). Cocci and bacilli frequently form clusters or linear filaments, and may have bacterial flagella. Relatively few species of bacteria cause infection. Hundreds of species of non-pathogenic bacteria live on the human body and within the gastrointestinal (GI) tract. Those in the GI tract constitute a person's gut fauna, and are biologically critical to humans.

Photosynthetic bacteria contain chlorophyll and release oxygen during photosynthesis. Some bacteria are *obligate aerobes* (require O_2 for metabolism) and others are *facultative anaerobes* (indifferent to O_2 for metabolism). Some are *obligate anaerobes* (oxygen may poison them). Most bacteria are heterotrophic *saprophytes*, which secrete enzymes to break down surrounding organic molecules into absorbable compounds.

Archaea are adapted to a limited range of extreme conditions. The cell walls of Archaea lack peptidoglycan (characteristic of bacteria). Archaea have distinctive RNAs and RNA polymerase enzymes. They include methanogens, typically found in swamps and marshes, and thermoacidophiles, found in acid hot springs, acidic soil, and deep oceanic volcanic vents.

Methanogens exist in oxygen-free environments and subsist on simple compounds such as CO_2, acetate, or methanol. As their name implies, Methanogens produce methane gas as a byproduct of metabolism. These organisms are typically found in organic-rich mud and sludge that often contain fecal wastes.

Thermoacidophiles are resistant to hot temperatures and high acid concentrations. The cell membrane of these organisms contains high amounts of saturated fats, and their enzymes and other proteins are able to withstand extreme conditions without denaturation. These microscopic organisms thrive in most hot springs and hot, acid soils.

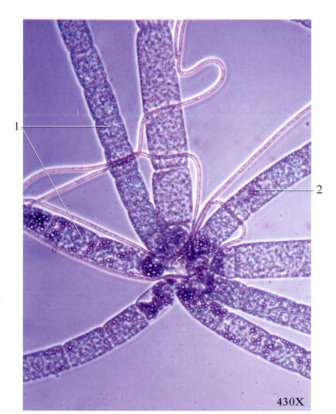

Figure 3.1 *Thiothrix* sp., a genus of bacteria that forms sulfur granules in its cytoplasm. These organisms obtain energy from oxidation of H_2S.
1. Filaments   2. Sulfur granules

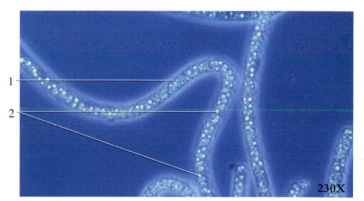

Figure 3.2 A filament of *Thiothrix* sp. with sulfur granules in its cytoplasm.
1. Filament   2. Sulfur granules

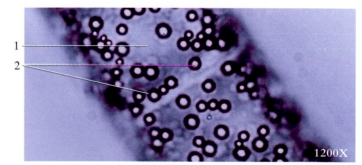

Figure 3.3 A magnified *Thiothrix* sp. filament with sulfur granules in its cytoplasm.
1. Cytoplasm   2. Sulfur granules

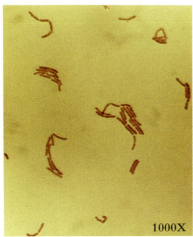

1000X

Figure 3.4 The bacterium *Bacillus megaterium*. *Bacillus* is capable of producing endospores. This species of *Bacillus* generally remains in chains after it divides.

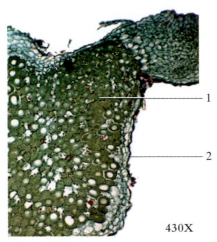

430X

Figure 3.5 A transverse section through the root of a clover stem showing intracellular nitrogen-fixing bacteria.
1. Cell with bacteria
2. Epidermis

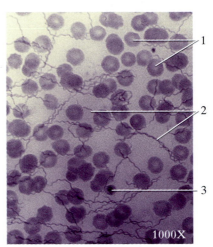

1000X

Figure 3.6 The spirochete, *Borella recurrentis*. Spirochetes are flexible rods twisted into helical shapes. This species causes relapsing fever.
1. Red blood cells
2. Spirochete
3. White blood cell

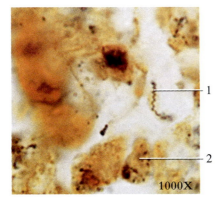

1000X

Figure 3.7 The spirochete *Treponema pallidum*. This species causes syphilis.
1. *Treponema pallidum*
2. White blood cell

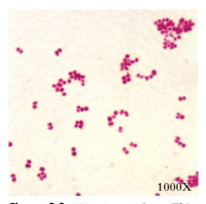

1000X

Figure 3.8 *Neisseria gonorrhoeae*. This is a diplococcus that causes gonorrhea.

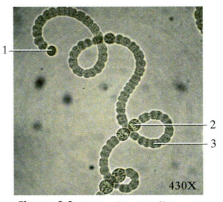

430X

Figure 3.9 An *Anabaena* sp. filament. This organism is a nitrogen-fixing cyanobacterium. Nitrogen fixation takes place within the heterocyst cells.
1. Heterocyst   3. Vegetative
2. Spore (akinete)   cell

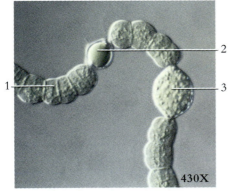

430X

Figure 3.10 An *Anabaena* sp. filament. This is a nitrogen-fixing cyanobacterium. Nitrogen fixation takes place within the heterocyst cells.
1. Vegetative cell   3. Spore
2. Heterocyst

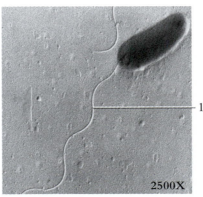

2500X

Figure 3.11 **The** flagellated bacterium, *Pseudomonas* sp.
1. Flagellum

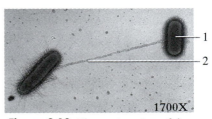

1700X

Figure 3.12 The conjugation of the bacterium *Escherichia coli*. By this process of conjugation, genetic material is transferred through the conjugation tube from one cell to the other allowing genetic recombination.
1. Bacterium   2. Conjugation tube

Table 3.1 Some Representatives of Bacteria and Archaea

Categories	Representative Genera
Bacteria	
Photosynthetic bacteria	
Cyanobacteria	*Anabaena, Oscillatoria, Spirulina, Nostoc*
Green bacteria	*Chlorobium*
Purple bacteria	*Rhodospirillum*
Gram-negative bacteria	*Proteus, Pseudomonas, Escherichia, Rhizobium, Neisseria*
Gram-positive bacteria	*Bacillus, Staphylococcus, Streptcoccus, Clostridium, Listeria*
Spirochetes	*Spirochaeta, Treponema*
Actinomycetes	*Streptomyces*
Rickettsias and Chlamydias	*Rickettsia, Chlamydia*
Mycoplasmas	*Mycoplasma*
Archaea	
Methanogens	*Halobacterium, Methanobacteria*
Thermoacidophiles	*Thermoplasma, Sulfolobus*

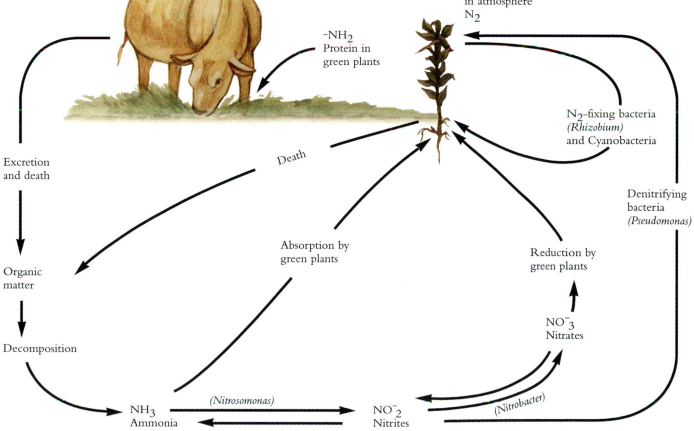

Figure 3.13 Few organisms have the ability to utilize atmospheric nitrogen. Nitrogen-fixing bacteria within the root nodules of legumes (and some free-living bacteria) provide a usable source of nitrogen to plants.

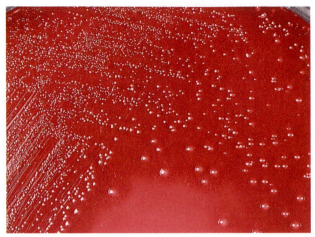

Figure 3.14 Colonies of *Streptococcus pyogenes* cultured on a sheep blood agar plate. *S. pyogenes* causes strep throat and rheumatic fever in humans. This agar plate is approximately 10 cm in diameter.

Figure 3.15 Cyanobacteria living in hot springs and hot streams, such as this 40 meter effluent from a geyser in Yellowstone National Park.
1. Mats of *Cyanophyta*

Figure 3.16 Cyanobacteria of several species growing in the effluent from a geyser. The different species are temperature-dependent and form the bands of color.

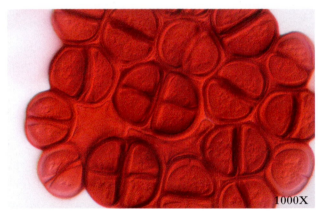

Figure 3.17 A magnified view of the cyanobacterium *Chroococcus* sp. shown with a red biological stain.

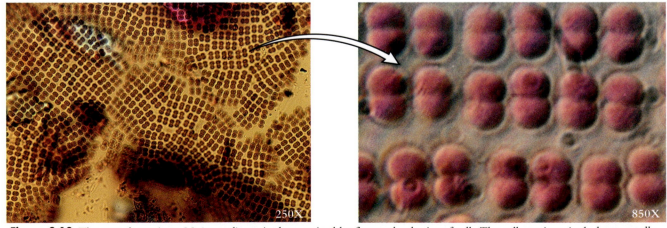

Figure 3.18 The cyanobacterium, *Merismopedia* sp., is characterized by flattened colonies of cells. The cells are in a single-layer, usually aligned into groups of two or four.

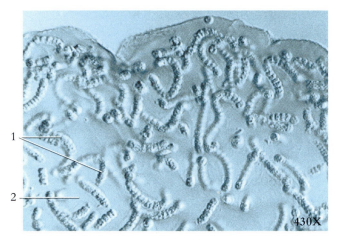

Figure 3.19 A colony of *Nostoc* sp. filaments. Individual filaments secrete mucilage, which forms a gelatinous matrix around all filaments.

1. Filaments   2. Gelatinous matrix

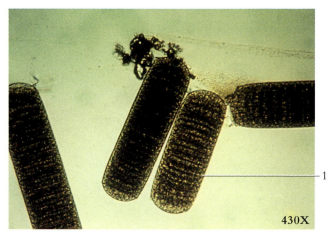

Figure 3.20 The filaments of *Oscillatoria* sp. The only way this cyanobacterium can reproduce is through fragmentation of a filament. Fragments are known as hormogonia.

1. Hormogonium

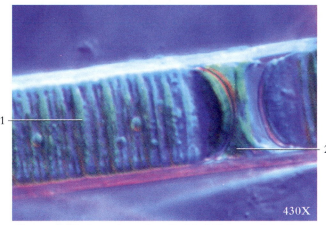

Figure 3.21 A portion of a cylindrical filament of *Oscillatoria* sp. This cyanobacterium is common in most aquatic habitats.

1. Filament segment   2. Separation disk
   (hormogonium)

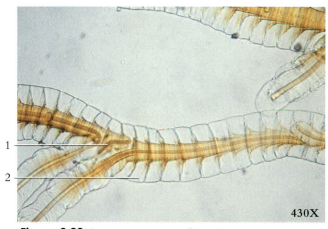

Figure 3.22 *Scytonema* sp., a cyanobacterium, common on moistened soil. Notice the falsely-branched filament typical of this genus. This species also demonstrates "winged" sheaths.

1. False branching   2. "Winged" sheath

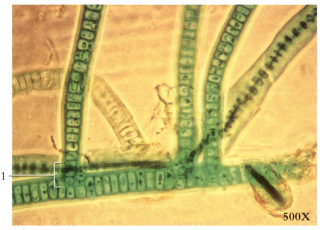

Figure 3.23 The cyanobacterium, *Stigonema* sp. This species has true-branched filaments caused from cell division in two separate planes.

1. True branching

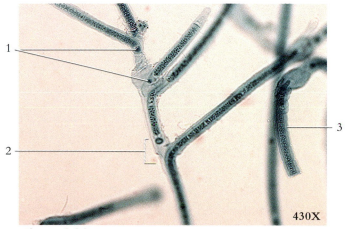

Figure 3.24 The cyanobacterium *Tolypothrix* sp., with a single false-branched filament.

1. Heterocysts   2. False branching   3. Sheath

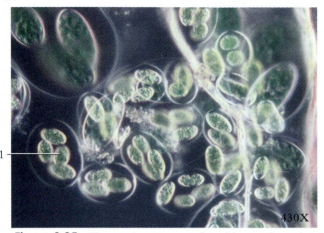

Figure 3.25 *Glaucocystis* sp., a green alga with cyanobacteria as endosymbionts.
  1. Cyanobacteria endosymbiont

Figure 3.26 *Microcoleus* sp., one of the most common cyanobacteria in and on soils throughout the world. It is characterized by several filaments in a common sheath.

Figure 3.27 A satellite image of a large lake. The circular pattern in the water is composed of dense growths of cyanobacteria.

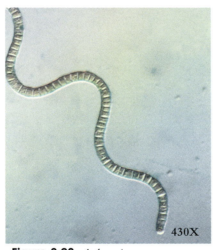

Figure 3.28 *Arthrospira* sp., a common cyanobacterium.

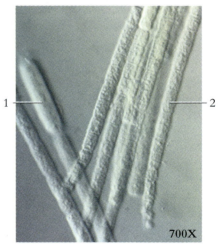

Figure 3.29 The cyanobacterium *Aphanizomenon* sp., common in nutrient-rich (often polluted) waters around the world.
  1. Spore (akinete)   2. Filament

Figure 3.30 A spring seep in Zion National Park, Utah.
  1. Mat of cyanobacteria.

Figure 3.31 A researcher examining cyanobacterial growths on soil in Canyonlands National Park, Utah.

Figure 3.32 A close-up photo of cryptobiotic soil crust. These crusts are composed of cyanobacteria, fungi, lichens, and other organisms.

Originally the 'Kingdom' Protista included only the unicellular eukaryotic organisms, but later was expanded to include all eukaryotes that are not fungi, animals, or land plants. Recent analyses that include a greater consideration of molecular data and broader representation of eukaryotic diversity reveals that the 'Protista' are more accurately depicted by placing them into several different kingdoms, or 'Eukaryote Super-groups' (see the Prelude to this atlas). This classification scheme is still being modified and has yet to be fully developed for lower taxonomic groups. Therefore, for convenience only, we maintain the historical terminology of Protista.

Protists are eukaryotic organisms that range in size from microscopic unicellular organisms to giant multicellular kelp. Protists are all composed of eukaryotic cells and, therefore, have a nucleus, mitochondria, endoplasmic reticulae, and Golgi complexes. Some contain chloroplasts. Most protists are capable of meiosis and sexual reproduction; these processes evolved a billion or more years ago and occur in nearly all complex plants and animals.

Protists are abundant in aquatic habitats, and are important constituents of plankton. Plankton are communities of organisms that drift passively or swim slowly in ponds, lakes, and oceans. Plankton are a major source of food for other aquatic organisms. Photosynthetic protists are the primary food producers in aquatic ecosystems.

The unicellular algal protists include microscopic aquatic organisms within the phyla Chrysophyta and Dinophyta. Chrysophyta are the yellow-green and golden-brown algae, and the diatoms. The cell wall of a diatom is composed largely of silica rather than cellulose. Some diatoms move in a slow, gliding way as cytoplasm glides through slits in the cell wall to propel the organism.

The Dinophyta are single-celled, algae-like organisms, the most important of which are the dinoflagellates. In most species of dinoflagellates, the cell wall is formed of armor-like plates of cellulose. Dinoflagellates are motile, having two flagella. Generally one encircles the organism in a transverse groove, and the other projects to the posterior.

Protozoa are also protists. They are small ($2\mu m - 100\mu m$), unicellular eukaryotic organisms that lack a cell wall. Movement of protozoa is either lacking or due to flagella, cilia, or pseudopodia of various sorts. In feeding upon other organisms or organic particles, they use simple diffusion, pinocytosis, active transport, or phagocytosis. Although most protozoa reproduce asexually, some species may also reproduce sexually during a portion of their life cycle. Most protozoa are harmless, although some are of immense clinical concern because they are parasitic and may cause human disease, including African sleeping sickness and malaria.

Table 4.1 Some Representatives of the Protista: Primarily Unicellular Organisms

	Taxa and Representative Kinds	Characteristics
Plant-like	**Chrysophyta**—diatoms and golden algae	Diatom cell walls composed of or impregnated with silica, often with two halves; plastids are often golden in Chrysophyta due to pigment composition
	Dinophyta—dinoflagellates	Two flagella in grooves of wall; brownish-gold plastids
Animal-like	**Rhizopoda**—amoebas	Cytoskeleton of microtubules and microfilaments; amoeboid locomotion
	Apicomplexa—sporozoa and Plasmodia	Lack locomotor capabilities and contractile vacuoles; mostly parasitic
	Sarcomastigophora—protozoa	Use flagella or pseudopodia to locomote; mostly parasitic
	Euglenophyta—euglenoids	Flagellates containing chloroplasts, lacking typical cell walls
	Ciliophora—ciliates and paramecia	Use cilia to move and feed

Figure 4.1 An illustration of *Amoeba proteus*, a fresh-water protozoan.

Phylum Chrysophyta - diatoms and golden algae

Figure 4.2 *Biddulphia*, a diatom forming colonies. These cells are beginning cell division.

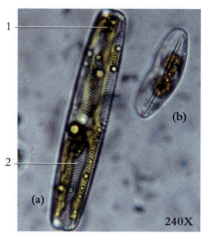

Figure 4.3 Live specimens of pennate (bilaterally symmetrical) diatoms. (a) *Navicula*, and (b) *Cymbella*.
1. Chloroplast   2. Striae

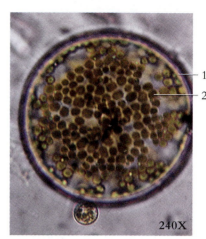

Figure 4.4 *Hyalodiscus,* a centric (radially symmetrical) diatom, from a freshwater spring in Nevada.
1. Silica cell wall   2. Chloroplasts

Figure 4.5 *Epithemia*, a distinctive pennate freshwater diatom.

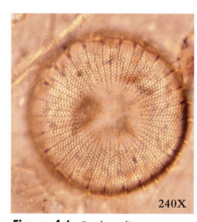

Figure 4.6 *Stephanodiscus*, a centric diatom.

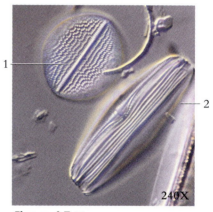

Figure 4.7 Two common freshwater diatoms.
1. *Cocconeis*   2. *Amphora*

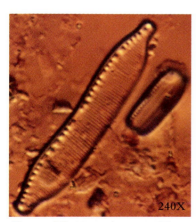

Figure 4.8 *Hantzschia*, one of the most common soil diatoms.

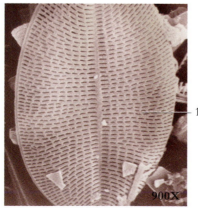

Figure 4.9 A scanning electron micrograph of *Cocconeis*, a common freshwater diatom.
1. Striae containing pores, or punctae, in the frustule (silicon cell wall).

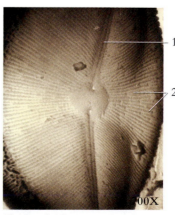

Figure 4.10 A scanning electron micrograph of the diatom *Achnanthes flexella*.
1. Raphe   2. Striae

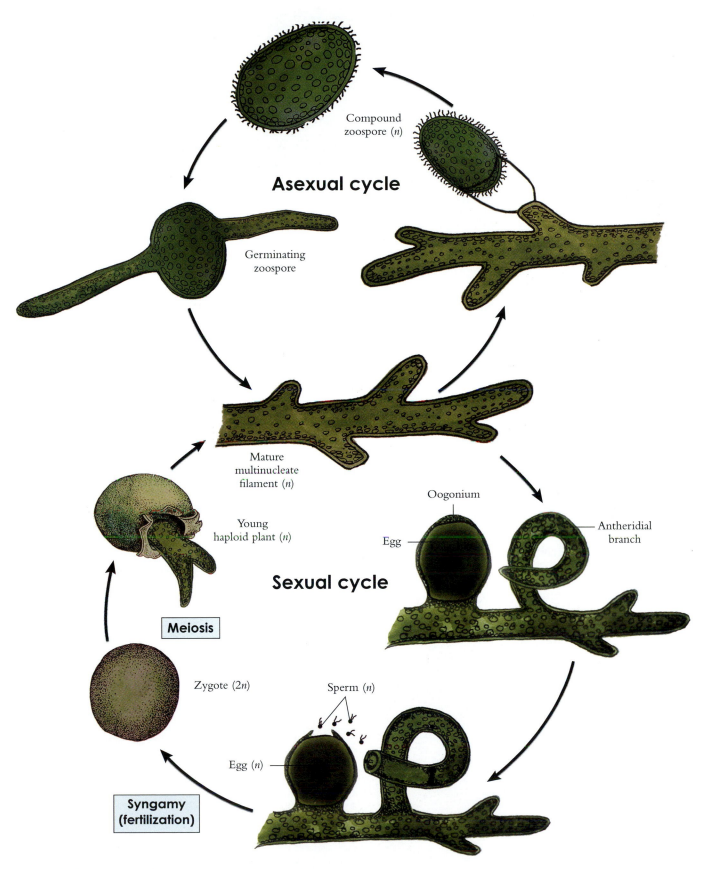

Figure 4.11 The life cycle of the "water felt alga," *Vaucheria*.

700X

Figure 4.12 A filament with immature gametangia of the "water felt" alga, *Vaucheria*. *Vaucheria* is a chrysophyte that is widespread in freshwater and marine habitats. It is also found in the mud of brackish areas that periodically become submerged and then exposed to air.
1. Antheridium
2. Developing oogonium

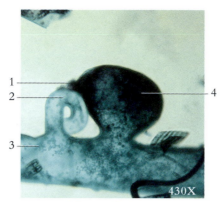

430X

Figure 4.13 A *Vaucheria*, with mature gametangia.
1. Fertilization pore
2. Antheridium
3. Chloroplasts
4. Developing oogonium

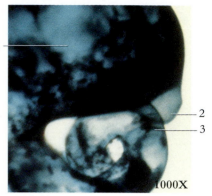

1000X

Figure 4.14 A *Vaucheria*, with mature gametangia.
1. Oogonium
2. Fertilization pore
3. Antheridium

Phylum Dinophyta (Pyrrhophyta) - dinoflagellates

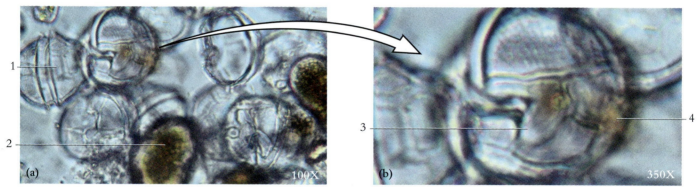

(a) 100X   (b) 350X

Figure 4.15 The dinoflagellates, *Peridinium*. (a) Some organisms are living; (b) others are dead and have lost their cytoplasm and consist of resistant cell walls.
1. Dead dinoflagellate
2. Living dinoflagellate
3. Cellulose plate
4. Remnant of cytoplasm

Figure 4.16 A giant clam with bluish coloration due to endosymbiont dinoflagellates.

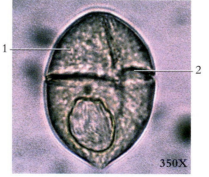

350X

Figure 4.17 A photomicrograph of *Peridinium* sp. The cell wall of many dinoflagellates is composed of overlapping plates of cellulose.
1. Transverse groove
2. Wall of cellulose plates

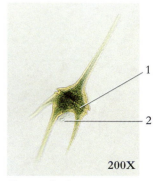

200X

Figure 4.18 *Ceratium* sp., a common freshwater dinoflagellate.
1. Transverse groove
2. Trailing flagellum

Phylum Rhizopoda - amoebas

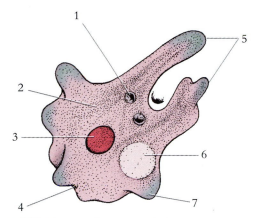

Figure 4.19 The *Amoeba proteus,* a freshwater protozoan that moves by forming cytoplasmic extensions called pseudopodia.

1. Food vacuole
2. Endoplasm
3. Nucleus
4. Cell membrane
5. Pseudopodia
6. Contractile vacuole
7. Ectoplasm

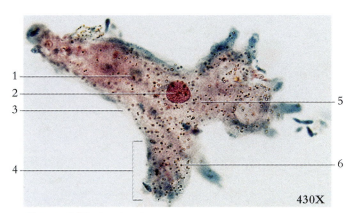

Figure 4.20 An *Amoeba proteus.*

| 1. Food vacuole | 3. Cell membrane | 5. Endoplasm |
| 2. Nucleus | 4. Pseudopodia | 6. Ectoplasm |

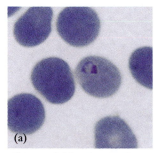

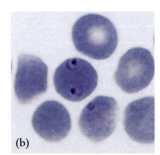

Figure 4.21 The protozoan *Entamoeba histolytica,* the causative agent of amebic dysentery, a disease most common in areas with poor sanitation. (a) A trophozoite, and (b) a cyst.

Phylum Apicomplexa - sporozoans and plasmodia

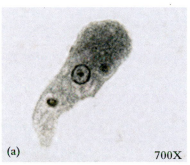
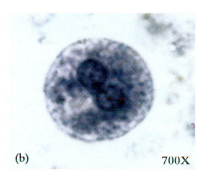
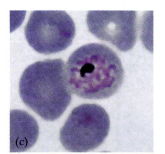
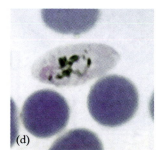

Figure 4.22 The protozoan *Plasmodium falciparum* causes malaria, which is transmitted by the female *Anopheles* mosquito. (a) The ring stage in a red blood cell, (b) a double infection, (c) a developing schizont, and (d) a gametocyte. All 200X

Phylum Sarcomastigophora - flagellated protozoans

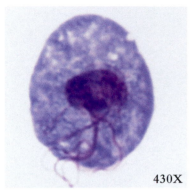

430X

Figure 4.23 The protozoan *Trichomonas vaginalis*, the causative agent of trichomoniasis. Trichomoniasis is an inflammation of the genitourinary mucosal surfaces—the urethra, vulva, vagina, and cervix in females and the urethra, prostate, and seminal vesicles in males.

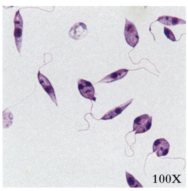

100X

Figure 4.24 The protozoan *Leishmania donovani*, the causative agent of visceral leishmaniasis, or kala-azar disease, in humans. The sandfly is the infectious host of this disease.

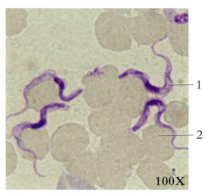

100X

Figure 4.25 The flagellated protozoan *Trypanosoma brucei*, the causative agent of African trypanosomiasis, or African sleeping sickness. The tsetse fly is the infectious host of this disease in humans.
1. *Trypanosoma brucei*
2. Red blood cell

Phylum Euglenophyta - euglenoids

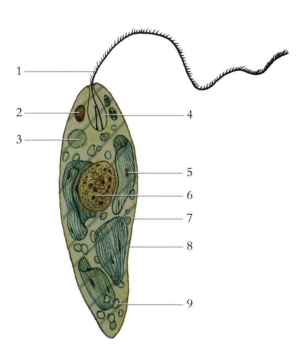

Figure 4.26 A diagram of *Euglena*, which contains flagellates that contain chloroplasts. They are freshwater organisms that have a flexible pellicle rather than a rigid cell wall.

1. Long flagellum
2. Photoreceptor (eye spot)
3. Contractile vacuole
4. Reservoir
5. Chloroplast
6. Nucleus
7. Pellicle
8. Cell membrane
9. Paramylon granule

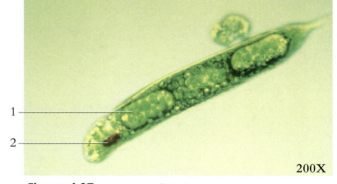

200X

Figure 4.27 A species of *Euglena*.
1. Paramylon body   2. Photoreceptor

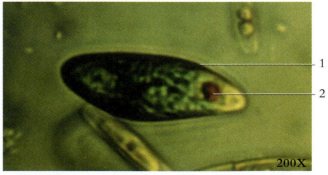

200X

Figure 4.28 A species of *Euglena* from a brackish lake in New Mexico.
1. Pellicle   2. Photoreceptor

Phylum Ciliophora - ciliates and paramecia

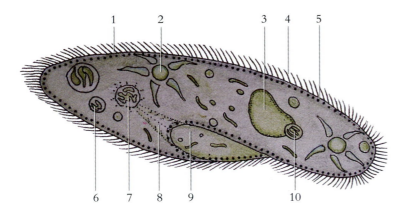

Figure 4.29 A *Paramecium caudatum,* a ciliated protozoan. The poisonous trichocysts of these unicellular organisms are used for defense and capturing prey.

1. Pellicle
2. Contractile vacuole
3. Macronucleus
4. Cilia
5. Trichocyst
6. Food vacuole

7. Forming food vacuole
8. Gullet
9. Oral cavity
10. Micronucleus

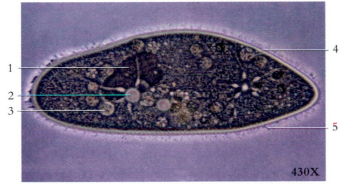

430X

Figure 4.30 A *Paramecium caudatum,* a ciliated protozoan.

1. Macronucleus
2. Contractile vacuole
3. Micronucleus

4. Pellicle
5. Cilia

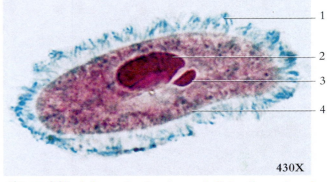

430X

Figure 4.31 A *Paramecium busaria,* a unicellular, slipper-shaped organism. Paramecia are usually common in ponds containing decaying organic matter.

1. Cilia
2. Macronucleus

3. Micronucleus
4. Pellicle

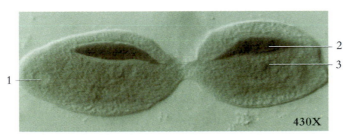

430X

Figure 4.32 A *Paramecium* in fission.

1. Contractile vacuole
2. Macronucleus

3. Micronucleus

430X

Figure 4.33 A *Paramecium* in conjugation.

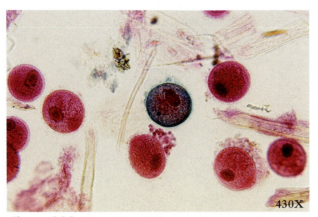

Figure 4.34 *Balantidium coli*, the causative agent of balantidiasis. Cysts in sewage-contaminated water are the infective form.

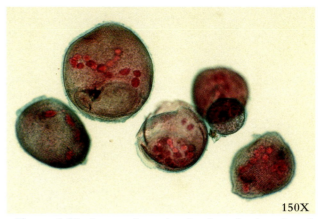

Figure 4.35 *Stentor*, a free-swimming form that has adopted an oval shape.

Table 4.2 Some Representatives of Protista: Primarily Multicellular Organisms

Phylum and Representative Kinds	Characteristics
Algae	
Phylum Chlorophyta—green algae	Unicellular, colonial, filamentous and multicellular plate-like forms; mosly freshwater; reproduce asexually and sexually
Phylum Phaeophyta—brown algae, giant kelp	Multicellular, mostly marine often in the intertidal zone; most with alternation of generations
Phylum Rhodophyta—red algae	Multicellular, mostly marine; sexual reproduction but with no flagellated cells; alternation of generations common
Protists Resembling Fungi	
Phylum Myxomycota—plasmodial slime molds	Multinucleated continuum of cytoplasm without cell membranes; amoeboid plasmodium during feeding stage; produce asexual fruiting bodies; gametes produced by meiosis
Phylum Dictyosteliomycota—cellular slime molds	Solitary cells during feeding stage; cells aggregate when food is scarce; produce asexual fruiting bodies
Phylum Oomycota—water molds, white rusts and downy mildews	Decomposers or parasitic forms; walls of cellulose, dispersal by non-motile spores or flagellated zoospores, gametes produced by meiosis

Phylum Chlorophyta - green algae

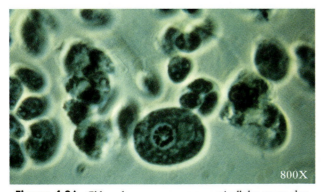

Figure 4.36 *Chlamydomonas*, a common unicellular green alga.

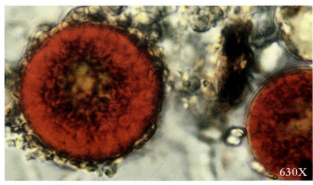

Figure 4.37 *Chlamydomonas nivalis*, the common snow alga.

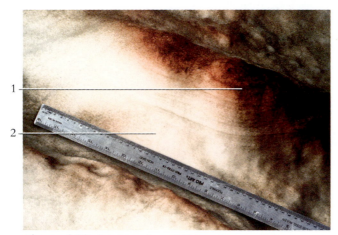

Figure 4.38 A habitat shot of *Chlamydomonas nivalis* creating "red snow."

  1. *Chlamydomonas nivalis*   2. Snow

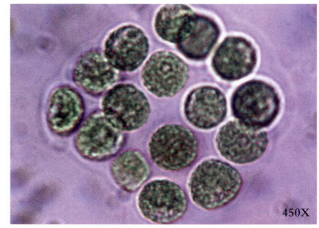

Figure 4.39 A *Gonium* sp. colony. *Gonium* sp. is a 16-celled flat colony of *Chlamydomonas*-like cells.

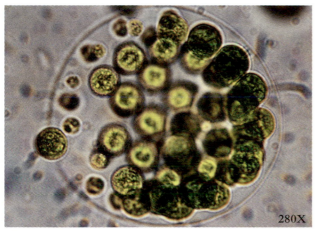

Figure 4.40 *Pleodorina* sp., is a multicellular colony (often 64-celled) relative of *Chlamydomonas* and *Volvox*.

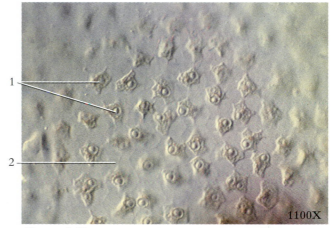

Figure 4.41 A close-up of the surface of *Volvox* sp. showing the interconnections between cells.

  1. Vegetative cells   2. Cytoplasmic connection between cells

Figure 4.42 A *Volvox* sp. Three separate organisms are shown in this photomicrograph, each containing daughter colonies of various ages.

  1. Immature daughter colony   3. Vegetative cells
  2. Daughter colonies

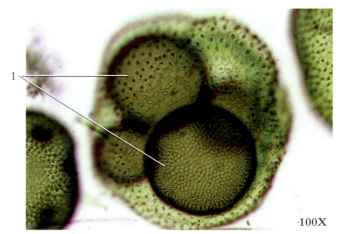

Figure 4.43 A single *Volvox* sp., organism with several large daughter colonies.

  1. Daughter colonies

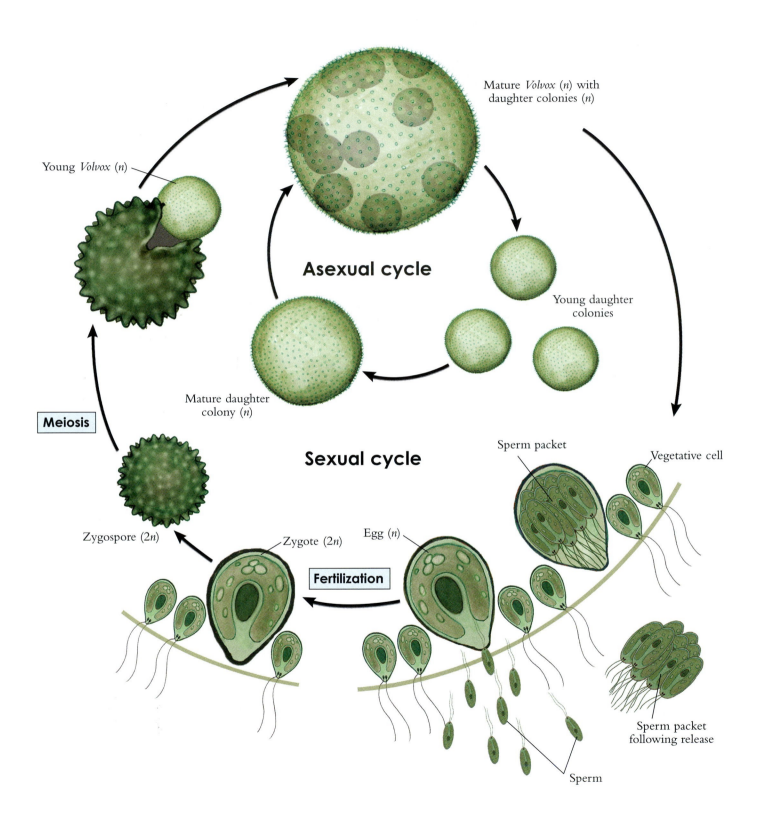

Mature *Volvox* (*n*) with
daughter colonies (*n*)

Young *Volvox* (*n*)

Asexual cycle

Young daughter
colonies

Meiosis

Mature daughter
colony (*n*)

Sexual cycle

Sperm packet

Vegetative cell

Zygospore (2*n*)

Zygote (2*n*)

Egg (*n*)

Fertilization

Sperm packet
following release

Sperm

Figure 4.44 The life cycle of *Volvox*, a common freshwater chlorophyte. *Volvox* is considered by
some to be a colony and by others to be a single, integrated plant.

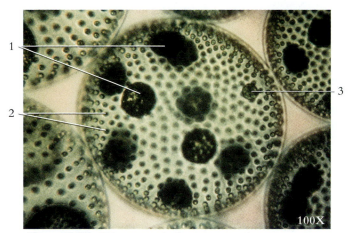

Figure 4.45 *Volvox* sp., a single mature specimen with several eggs and zygotes.
  1. Zygotes   2. Vegetative cells   3. Egg

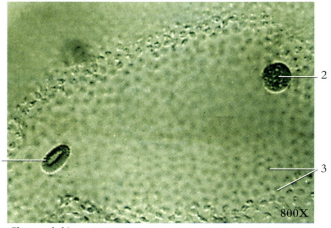

Figure 4.46 A single mature specimen of *Volvox* sp. This photomicrograph is a highly magnified view of a single organism showing gametes.
  1. Sperm packet   3. Vegetative cells
  2. Egg

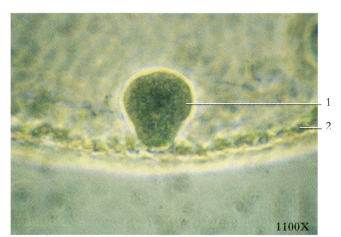

Figure 4.47 *Volvox* sp., showing a prominent egg at the edge of the organism. This egg will be fertilized to develop a zygote and then a zygospore.
  1. Egg   2. Vegetative cells

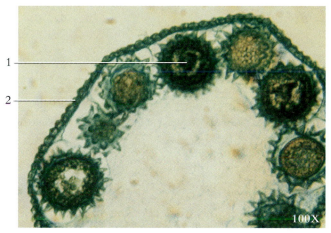

Figure 4.48 *Volvox* sp., a single mature organism with zygospores.
  1. Zygospore
  2. Vegetative cells

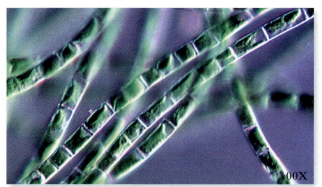

Figure 4.49 A live specimens of *Ulothrix* sp., an unbranched, filamentous green alga.

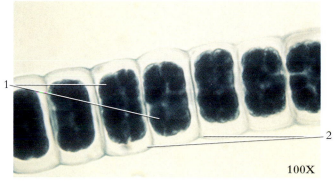

Figure 4.50 *Ulothrix* sp., an unbranched, filamentous green alga.
  1. Zoospores   2. Individual cells (known as sporangia when they produce spores)

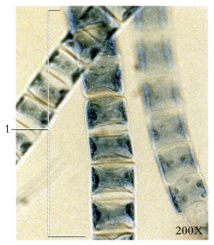

Vegetative stained filament

200X

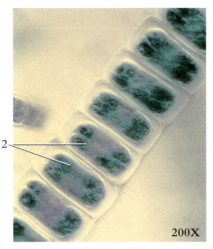

Stained filament with zoospores

200X

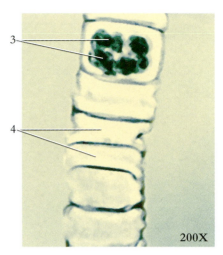

Empty filament, after zoospores have been released

200X

Figure 4.51 The production and release of zoospores in the green alga *Ulothrix* sp.
1. Filament   2. Young zoospores   3. Mature zoospores   4. Empty cells following zoospore release

Figure 4.52 A *Stigeocolonium* sp., a close relative of *Ulothrix*, showing a branched thallus.

Figure 4.53 A *Draparnaldia* sp., a relative of *Ulothrix*, showing different cell sizes in the thallus and a characteristic branching pattern.

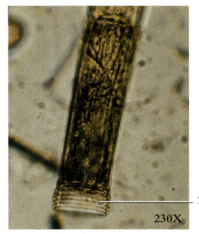

230X

Figure 4.54 The *Oedogonium* sp. with distinct "apical caps" that accrue from cell division in this genus.
1. Apical caps

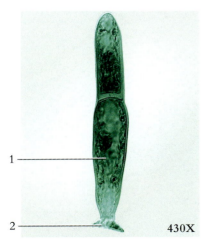

430X

Figure 4.55 A young filament of *Oedogonium* sp.
1. Basal cell
2. Holdfast

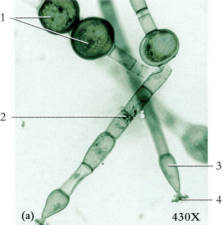

(a)   430X

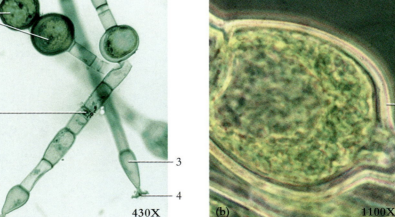

(b)   1100X

Figure 4.56 (a) *Oedogonium* sp., a filamentous, unbranched, green alga.
(b) Close-up of an oogonium.
1. Oogonia   3. Basal holdfast cell
2. Antheridium   4. Holdfast

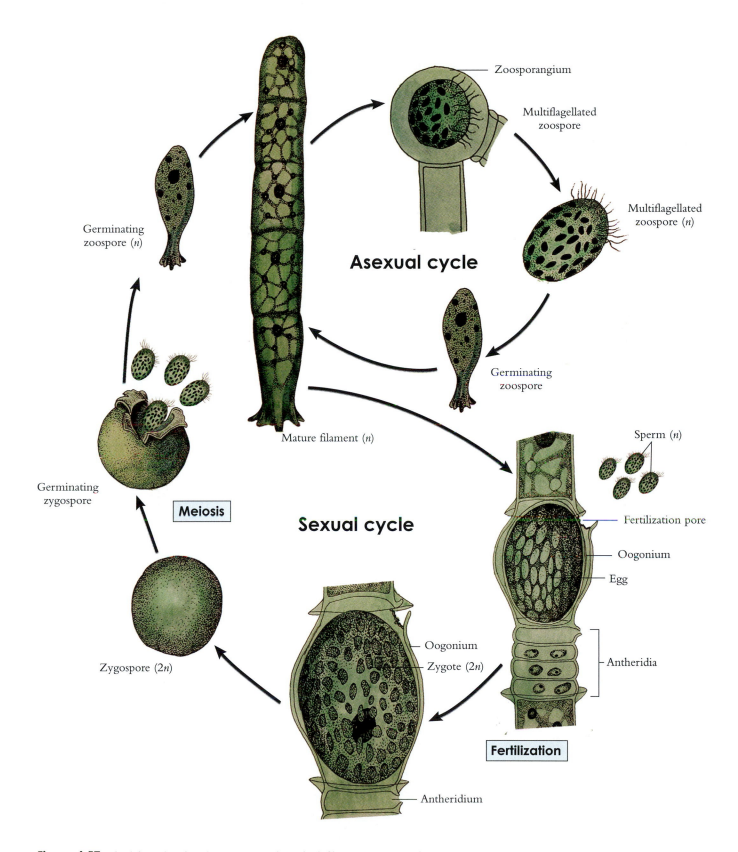

Figure 4.57 The life cycle of *Oedogonium*, an unbranched, filamentous green alga.

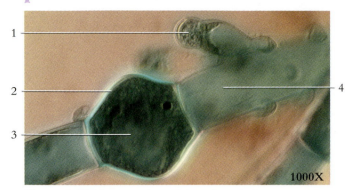

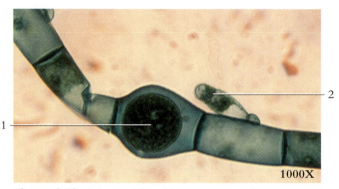

Figure 4.58 The oogonium of the unbranched green alga, *Oedogonium* sp.

1. Dwarf male filament   3. Developing egg
2. Oogonium   4. Vegetative cell

Figure 4.59 *An* oogonium with mature egg and dwarf male filament.

1. Egg
2. Dwarf male filament

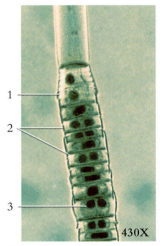

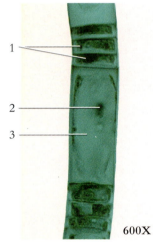

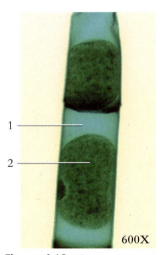

Figure 4.60 A filament of the green alga, *Oedogonium* sp.

1. Annular scars from cell division
2. Antheridia
3. Sperm

Figure 4.61 The green alga, *Oedogonium* sp., showing antheridia between vegetative cells.

1. Sperm within antheridia
2. Nucleus of vegetative cell
3. Vegetative cell

Figure 4.62 The zoosporangium of the unbranched green alga, *Oedogonium* sp.

1. Zoosporangium
2. Zoospore

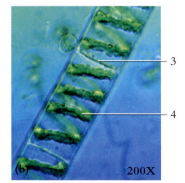

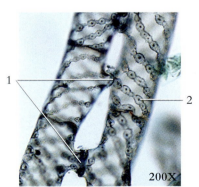

Figure 4.63 A species of *Spirogyra* are filamentous green algae commonly found in green masses on the surfaces of ponds and streams. Their chloroplasts are arranged as a spiral within the cell. (a) Several cells comprise a filament. (b) A magnified view of a single filament composed of several cells.

1. Single cell   2. Filaments   3. Cell wall   4. Chloroplast

Figure 4.64 The filaments of *Spirogyra* sp. showing initial contact of conjugation tubes.

1. Conjugation tube
2. Pyrenoid in chloroplast

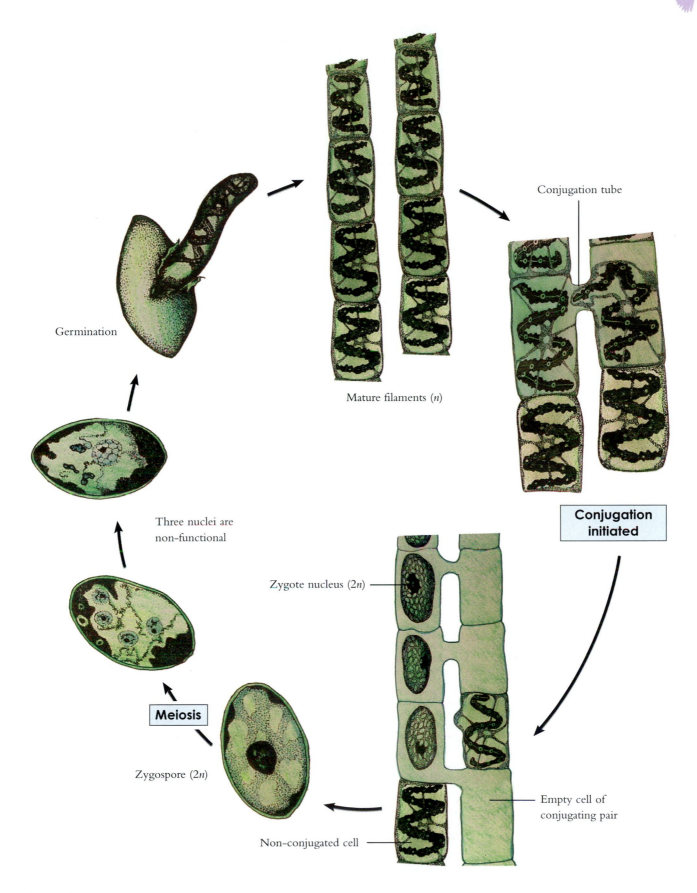

Conjugation tube

Mature filaments (*n*)

Conjugation initiated

Germination

Three nuclei are non-functional

Zygote nucleus (2*n*)

Meiosis

Zygospore (2*n*)

Empty cell of conjugating pair

Non-conjugated cell

Figure 4.65 The life cycle of *Spirogyra*, a common freshwater green alga.

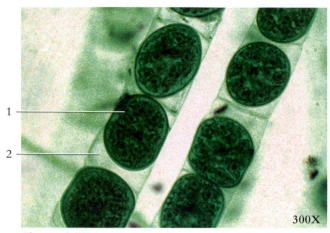

Figure 4.66 Two filaments of *Spirogyra* sp. with aplanospores.
1. Aplanospore   2. Cell wall

Figure 4.67 *Spirogyra* sp. in a small freshwater pond.

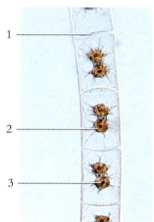

Figure 4.68 *Zygnema* sp. filament showing the star-shaped chloroplasts.
1. Cell wall
2. Chloroplast
3. Pyrenoid

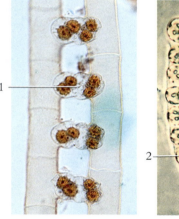

Figure 4.69 *Zygnema* sp. showing two locations of fertilization, (a) in the conjugation tube and (b) in cells of one of the conjugating filaments.
1. Fusing gametes   3. Cell wall
2. Zygote   4. Conjugation tube

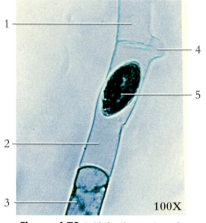

Figure 4.70 Self–fertile species of *Spirogyra* sp. A gamete has migrated from the upper cell to form a zygote in the lower cell.
1. Upper cell   4. Conjugation
2. Lower cell   tube
3. Chloroplast   5. Zygote

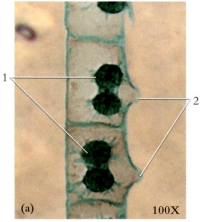

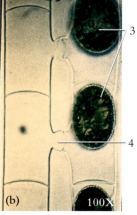

Figure 4.71 *Zygonema* sp. undergoing conjugation. (a) The filament is just forming conjugation tubes; and (b) two conjugated filaments.
1. Developing gametes   3. Zygotes
2. Developing conjugation tubes   4. Conjugation tube

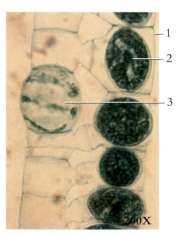

Figure 4.72 Conjugation in *Spirogyra* sp.
1. Cell bearing zygote
2. Zygote
3. Cell that did not conjugate

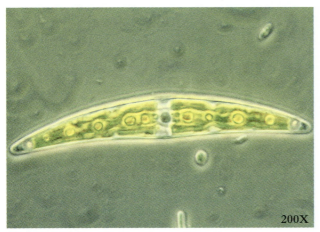

Figure 4.73 Desmid *Closterium* sp. Desmids are unicellular, freshwater chlorophyta, which reproduce sexually by conjugation.

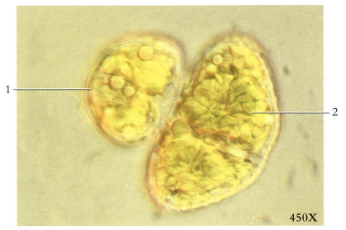

Figure 4.74 *Cosmorium* sp., a desmid, soon after cell division forming a new semi-cell.
1. New semi-cell   2. Dividing cell

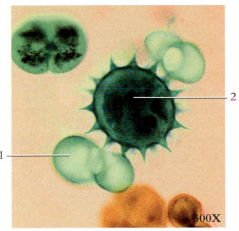

Figure 4.75 Zygospore of the desmid *Cosmarium* sp.
1. Empty cell that has been involved in conjugation
2. Zygospore

Figure 4.76 Desmid *Micrasterias* sp.

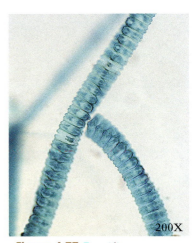

Figure 4.77 *Desmidium* sp., a filamentous (colonial) desmid.

Figure 4.78 Sea lettuce, *Ulva* sp., which lives as a flat membranous chlorophyte in marine environments.

Figure 4.79 Magnified view of the surface of *Enteromorpha intestinalis*. *Enteromorpha is* closely related to *Ulva.*

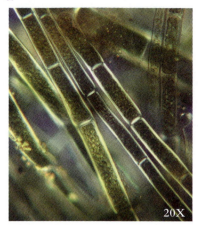

Figure 4.80 The filaments of *Cladophora* sp. This member of class Ulvophyceae is found in both freshwater and marine habitats.

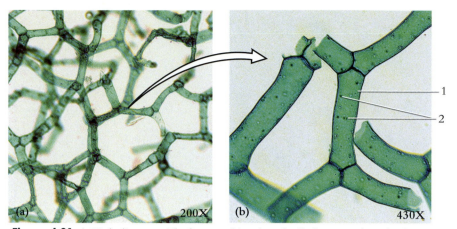

Figure 4.81 A *Hydrodictyon* sp.. The large, multinucleated cells form net-shaped colonies.
1. Individual cell
2. Nuclei of cell

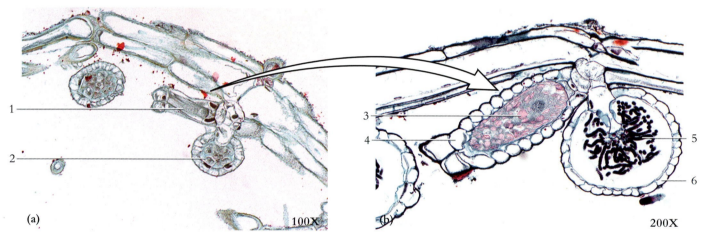

Figure 4.82 (a) *Chara* sp. inhabits marshes or shallow, temperate lakes, showing characteristic gametangia. (b) A magnified view of the gametangia.
1. Oogonium
2. Antheridium
3. Egg
4. Oogonium
5. Sperm-producing cells (filaments)
6. Antheridium

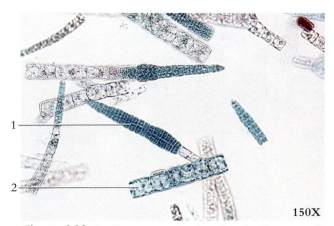

Figure 4.83 An *Ectocarpus* sp. showing pleurolocular sporangia.
1. Pleurolocular sporoangium
2. Filament of cells

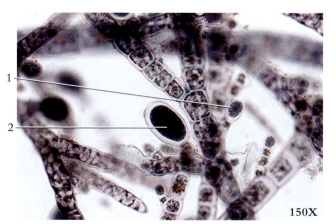

Figure 4.84 An *Ectocarpus* sp. showing unilocular sporangia.
1. Immature unilocular sporangium
2. Mature unilocular sporangium

Phylum Phaeophyta - brown algae and giant kelp

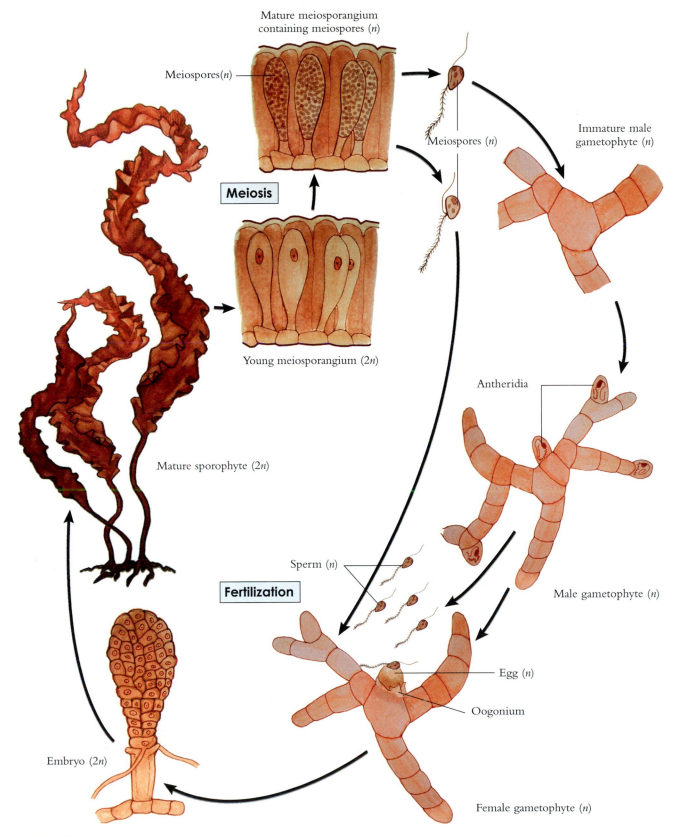

Mature meiosporangium
containing meiospores (n)

Meiospores(n)

Meiospores (n)

Immature male
gametophyte (n)

Meiosis

Young meiosporangium (2n)

Antheridia

Mature sporophyte (2n)

Sperm (n)

Male gametophyte (n)

Fertilization

Egg (n)

Oogonium

Embryo (2n)

Female gametophyte (n)

Figure 4.85 The life cycle of *Laminaria*, a common kelp.

Figure 4.86 Rocky coast of southern Alaska showing dense growths of the brown alga, *Fucus* sp.

Figure 4.87 "Sea palm," *Postelsia palmaeformis*, a common brown alga found on the western coast of North America.

Macrocystis sp.

Macrocystis sp.

Macrocystis sp.

Egregia sp.

Figure 4.88 Some examples of brown algae, Phaeophyta. These large species are commonly known as kelps.

  1. Blade   2. Float (air-filled bladder)   3. Stipe

Figure 4.89 The kelp, *Laminaria* sp., one of the common "sea weeds" found along many rocky coasts.

Figure 4.90 A tidal pool with green, brown, and red algae.

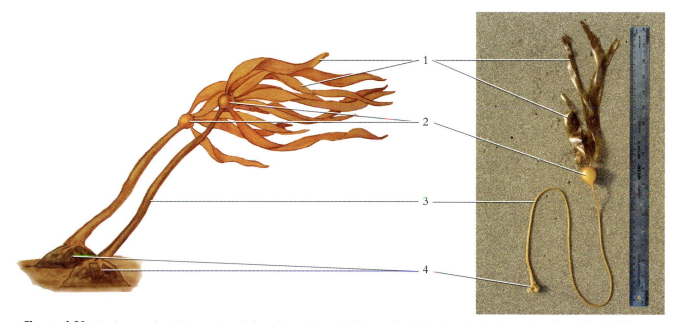

Figure 4.91 The brown alga, *Nereocystis* sp. It has a long stipe and photosynthetic laminae attached to a large float. The holdfast anchors the alga to the ocean floor. This alga and others can grow to lengths of several meters.

1. Lamina   2. Floats (air-filled bladders)   3. Stipe   4. Holdfasts

Figure 4.92 *Sargassum* sp., a brown alga common in the Sargasso sea.

1. Floats   2. Blade   3. Stipe

Figure 4.93 A mixture of kelps washed onto shore to form "windrows" of *Phaeophyta* sp.

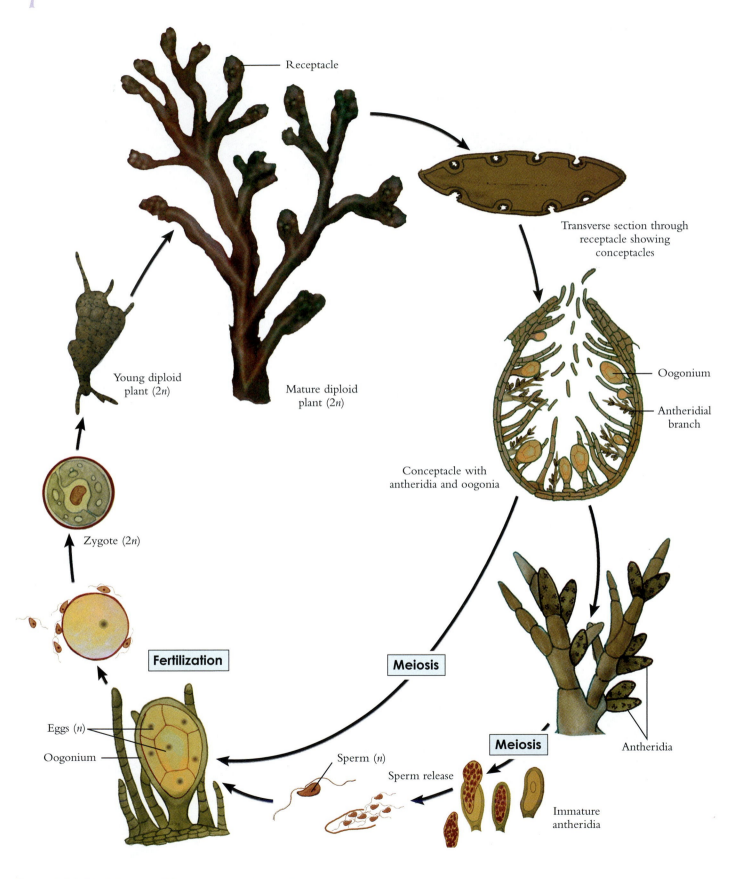

Receptacle

Transverse section through
receptacle showing
conceptacles

Oogonium

Antheridial
branch

Young diploid
plant (2n)

Mature diploid
plant (2n)

Conceptacle with
antheridia and oogonia

Zygote (2n)

Meiosis

Antheridia

Fertilization

Meiosis

Eggs (n)

Oogonium

Sperm (n)

Sperm release

Immature
antheridia

Figure 4.94 The life cycle of *Fucus*, a common brown alga.

Figure 4.95 (a) *Fucus* sp., a brown alga, commonly called rockweed. (b) An enlargement of a blade supporting the receptacles.

1. Blade
2. Receptacle
3. Conceptacles (spots) are chambers imbed-ded in the receptacles
4. Blade

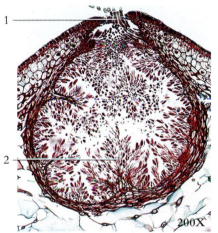

Figure 4.96 A conceptacle of *Fucus* sp.
1. Sterile paraphases
2. Antheridial branches

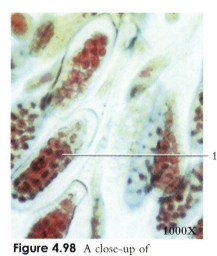

Figure 4.97 A close-up of antheridial branch of *Fucus* sp.
1. Antheridial branch
2. Antheridium

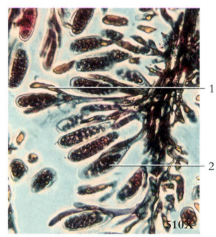

Figure 4.98 A close-up of antheridium of *Fucus* sp.
1. Sperm within antheridium

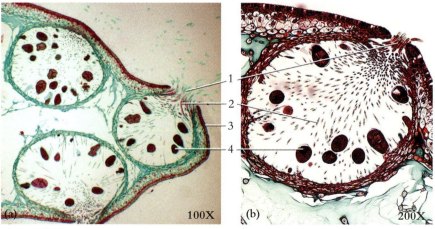

Figure 4.99 A section through a *Fucus* sp. recepticle. (a) Low magnification showing three conceptacles and (b) higher magnification of a single conceptacle with oogonia.
1. Ostiole
2. Paraphyses (sterile hairs)
3. Surface of receptacle
4. Oogonium

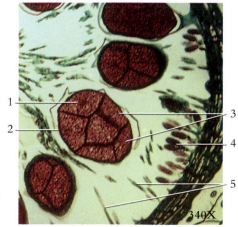

Figure 4.100 A close-up of the bisexual conceptacle of *Fucus* sp.
1. Nucleus of egg
2. Oogonium
3. Eggs
4. Antheridium
5. Paraphyses

Phylum Rhodophyta - red algae

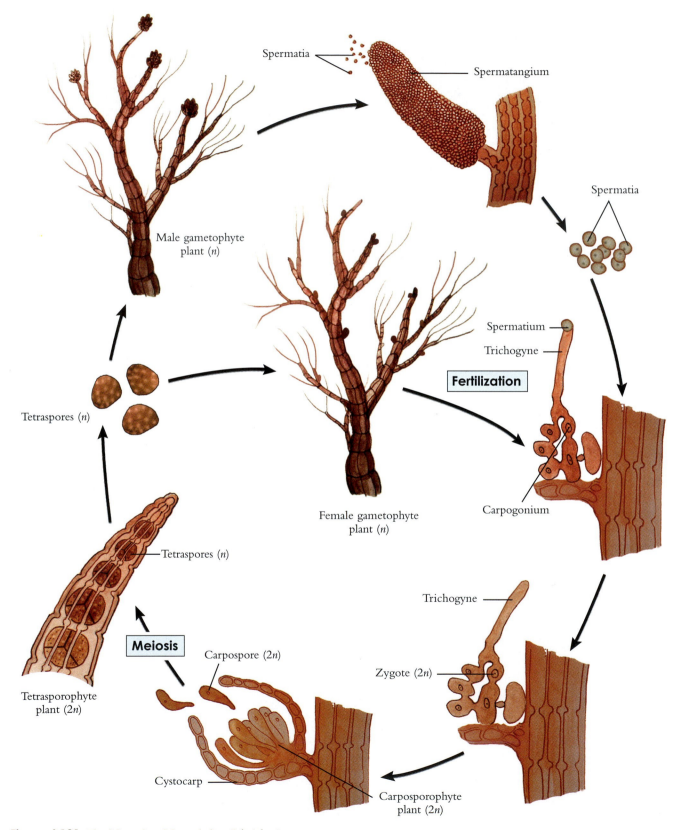

Figure 4.101 The life cycle of the red alga, *Polysiphonia*.

Figure 4.102 Intertidal zone showing a colony of red alga, *Bangia* sp.

Figure 4.103 Mature plant of the red alga, *Rhodymenia* sp.

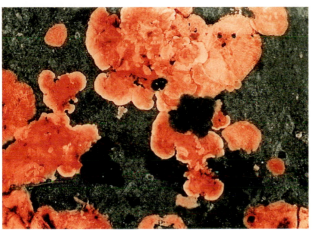

Figure 4.104 Small encrusting colonies of a species of red alga on a stone. The colonies shown are bright red and are only a few millimeters in size.

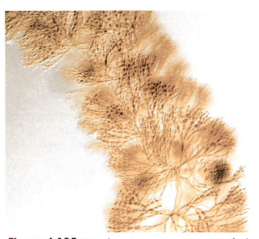

250X

Figure 4.105 *Batrachospermum* sp., a common freshwater red alga.

270X

Figure 4.106 *Audouinella* sp. is a freshwater member of Rhodophyta. This organism was collected from a coldwater spring.

Figure 4.107 Mature plant of the common red alga, *Polysiphonia* sp.

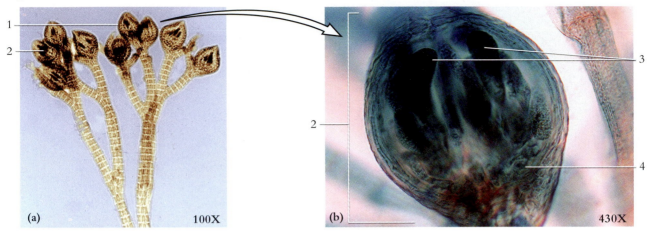

Figure 4.108 The red alga, *Polysiphonia* sp. It has alternation of three generations. (a) Female gametophyte with attached carposporophyte generation. (b) A close-up of the cystocarp plant.

1. Pericarp   2. Cystocarp   3. Carpospores   4. Carposporophyte

Figure 4.109 *Polysiphonia* sp., showing the release of carpospores.

1. Carpospores (2*n*)   2. Ruptured cystocarp

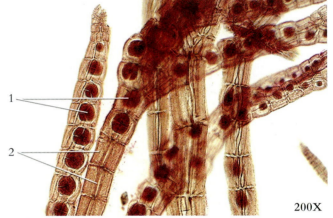

Figure 4.110 A tetrasporophyte generation of *Polysiphonia* sp. showing tetraspores (meiospores).

1. Tetraspores   2. Cells of tetrasporophyte plant

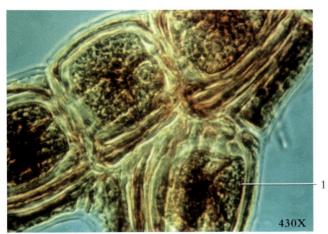

Figure 4.111 A close-up of tetrasporophyte plant of *Polysiphonia* sp.

1. Tetraspore (meiospore)

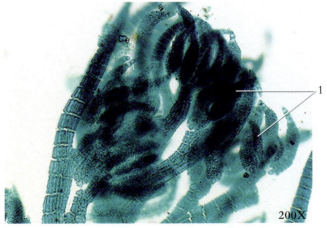

Figure 4.112 A male gametophyte plant of *Polysiphonia* sp., showing spermatangia (green stain).

1. Spermatangia with spermatia

Phylum Myxomycota (= Myxogastrida) - plasmodial slime molds

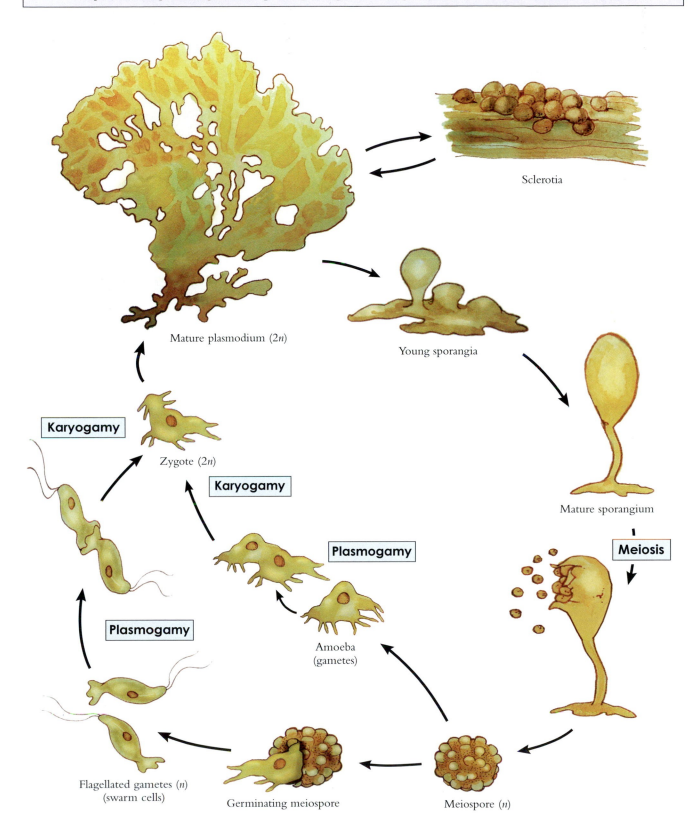

Sclerotia

Mature plasmodium (2*n*)

Young sporangia

Mature sporangium

Karyogamy

Zygote (2*n*)

Karyogamy

Meiosis

Plasmogamy

Amoeba (gametes)

Plasmogamy

Flagellated gametes (*n*) (swarm cells)

Germinating meiospore

Meiospore (*n*)

Figure 4.113 The life cycle of a plasmodial slime mold.

Figure 4.114 The sporangia of the slime mold
Comatricha typhoides.

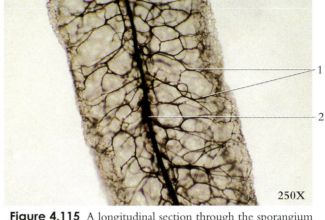

Figure 4.115 A longitudinal section through the sporangium
of *Stemonitis*.
1. Cellular filaments (capillitum)   2. Columella

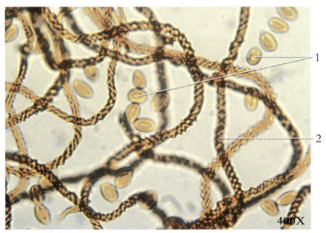

Figure 4.116 A close-up through the sporangium of
Stemonitis.
1. Spores   2. Capillitum

Figure 4.117 A *Physarum* plasmodium.

Figure 4.118 The sporangia of slime mold. These vary
considerably in size and shape. One species of *Lycogala*
is shown here.

Figure 4.119 A slime mold specimen from a high-
mountain locality.

Phylum Oomycota - water molds, white rusts, and downy mildews

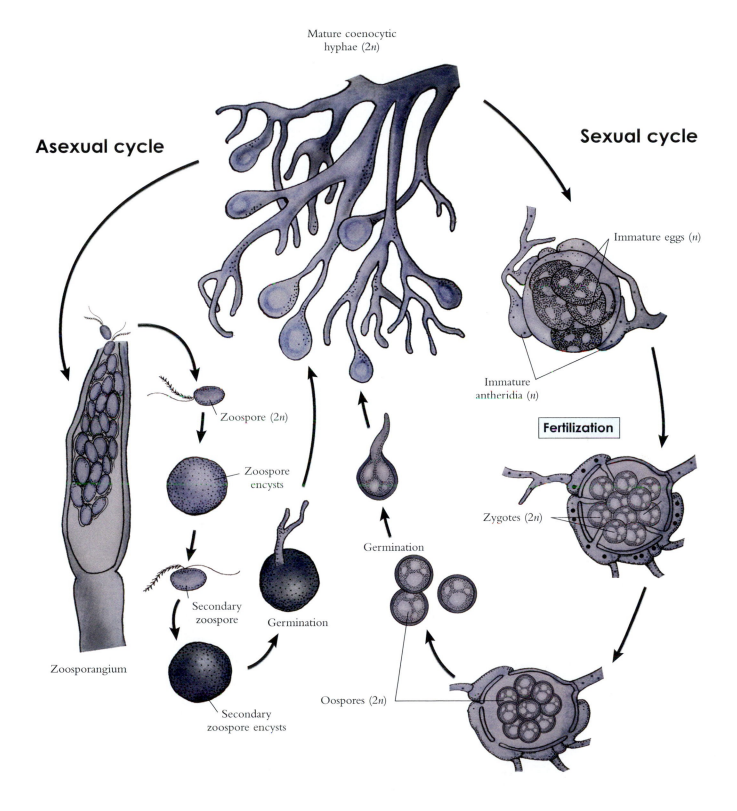

Mature coenocytic
hyphae (2n)

Asexual cycle

Sexual cycle

Immature eggs (n)

Immature
antheridia (n)

Fertilization

Zoospore (2n)

Zoospore
encysts

Zygotes (2n)

Germination

Secondary
zoospore   Germination

Zoosporangium

Germination

Oospores (2n)

Secondary
zoospore encysts

Figure 4.120 The life cycle of the water mold *Saprolegnia*.

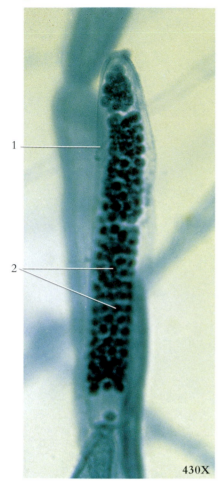

Figure 4.121 The zoosporangium of the water mold *Saprolegnia* sp.
1. Zoosporangium
2. Zoospores

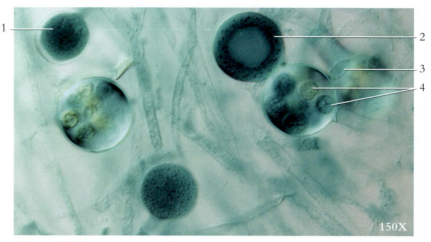

Figure 4.122 The oogonia of the water mold *Saprolegnia* sp.
1. Young oogonium   3. Young antheridium
2. Developing oogonium   4. Eggs

Figure 4.123 The water mold, *Saprolegnia* sp., showing a young oogonium before eggs have been formed.

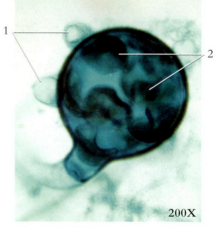

Figure 4.124 The oogonia of the water mold *Saprolegnia* sp.
1. Empty antheridia
2. Zygotes

Figure 4.125 The skin of this brown trout has been infected by the common water mold, *Saprolegnia* sp.

About 250,000 species of fungi are currently extant on Earth. All fungi are heterotrophs; they absorb nutrients through their cell walls and cell membranes. The kingdom Fungi includes the conjugation fungi, yeasts, mushrooms, toadstools, rusts, and lichens. Most are saprobes, absorbing nutrients from dead organic material. Some are parasitic, absorbing nutrients from living hosts. Fungi decompose organic material, helping to recycle nutrients essential for plant growth.

Except for the unicellular yeasts, fungi consist of elongated filaments called *hyphae*. Hyphae begin as cellular extensions of spores that branch as they grow to form a network of hyphae called a *mycelium*. Even the body of a mushroom consists of a mass of tightly packed hyphae attached to an underground mycelium. Fungi are nonmotile and reproduce by means of spores, produced sexually or asexually.

Many species of fungi are commercially important. Some are used as food, such as mushrooms; or in the production of foods, such as bread, cheese, beer, and wine. Other species are important in medicine, for example, in the production of the antibiotic penicillin. Many other species of fungi are of medical and economic concern because they cause plant and animal diseases and destroy crops and stored goods.

Table 5.1 Some Representatives of Fungi

Phyla and Representative Kinds	Characteristics
Zygomycota — conjugation fungi	Hyphae lack cross walls along filaments; sexual reproduction by conjugation
Ascomycota — yeasts, molds, morels, and truffles	Septate hyphae; reproductive structure contains meiospores within asci in fruiting bodies known as ascocarps (ascoma); asexual reproduction by budding or conidia
Basidiomycota — mushrooms, toadstools, rusts, and smuts	Septate hyphae; four meiospores produced externally on cells called basidia contained in basidiocarp (basidioma)
Lichens — not a phylum, but a symbiotic association of an alga and a fungus	Algal component (usually a green alga) provides food from photosynthesis; fungal component (usually an ascomycete) may provide anchorage, water retention, and/or nutrient absorbance

Figure 5.1 Two examples of common fungi. (a) A cup fungus, *Peziza*, a common ascomycete, and (b) oyster mushrooms, *Pleurotus*.

Phylum Zygomycota - conjugation fungi

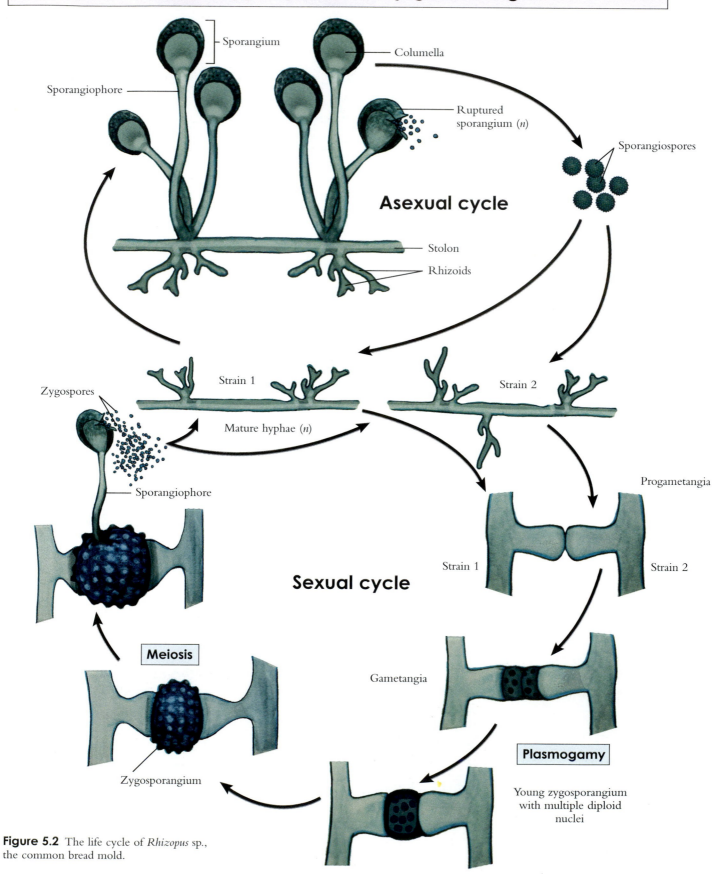

Figure 5.2 The life cycle of *Rhizopus* sp., the common bread mold.

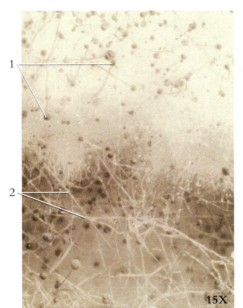

15X

Figure 5.3 A *Rhizopus* species growing on a slice of bread.
1. Sporangia
2. Hyphae (stolon)

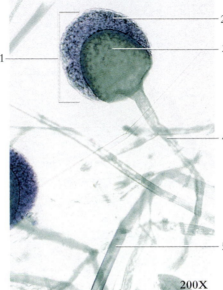

200X

Figure 5.4 A whole mount of the bread mold, *Rhizopus* sp.
1. Sporangium
2. Spores
3. Columella
4. Sporangiophore
5. Hyphae

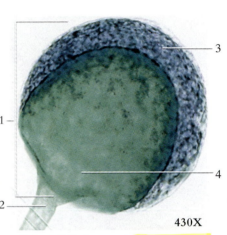
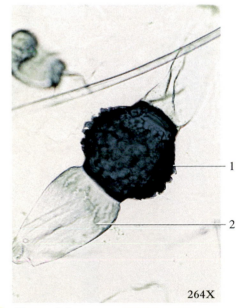

430X

Figure 5.5 A mature sporangium in the asexual reproductive cycle of the bread mold, *Rhizopus* sp.
1. Sporangium
2. Sporangiophore
3. Spores
4. Columella

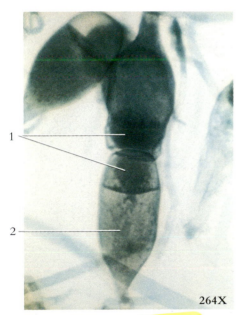

264X

Figure 5.6 A young gametangia of *Rhizopus* sp. contacting prior to plasmogamy.
1. Immature gametangia
2. Suspensor cell

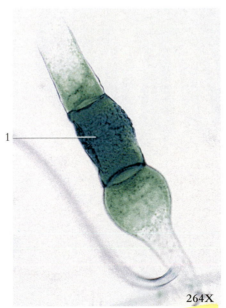

264X

Figure 5.7 An immature *Rhizopus* sp. zygospore following plasmogamy.
1. Immature zygosporangium

264X

Figure 5.8 A mature *Rhizopus* sp. zygospore.
1. Zygosporangium
2. Suspensor cell

Phylum Ascomycota -yeasts, molds, morels, and truffles

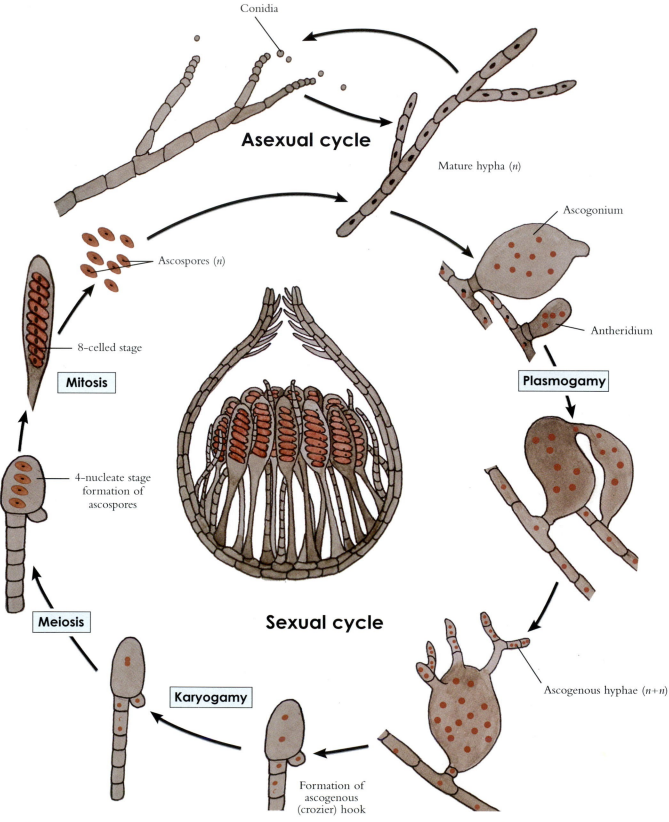

Conidia

Asexual cycle

Mature hypha (n)

Ascogonium

Ascospores (n)

Antheridium

8-celled stage

Mitosis

Plasmogamy

4-nucleate stage formation of ascospores

Meiosis

Ascogenous hyphae (n+n)

Sexual cycle

Karyogamy

Formation of ascogenous (crozier) hook

Figure 5.9 The life cycle of an ascomycete.

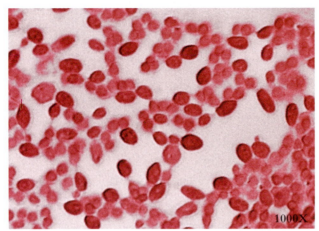

Figure 5.10 Baker's yeast, *Saccharomyces cerevisiae*. The ascospores of this unicellular ascomycete are characteristically spheroidal or ellipsoidal in shape.

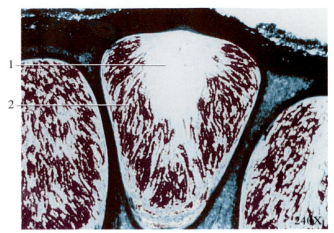

Figure 5.11 A close-up of the parasitic ascomycete, *Hypoxylon* sp., showing imbedded perithecia.
1. Perithecium   2. Hymenium

Figure 5.12 The parasitic ascomycete, *Dibotryon morbosum*, on a branch of a choke cherry, *Prunus virginiana*.
1. Fungus   2. Choke cherry stem

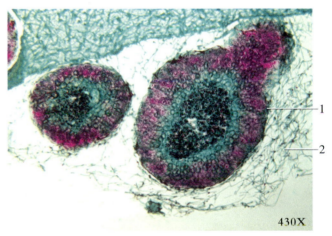

Figure 5.13 The cleistothecium (ascocarp or ascoma) of the ascomycete *Penicillium* sp.
1. Cleistothecium
2. Hyphae

Peziza repanda

Scutellinia scutellata

Morchella sp.

Helvella sp.

Figure 5.14 Fruiting bodies (ascocarps or ascoma) of common ascomycetes. *Peziza repanda* is a common woodland cup fungus. *Scutellinia scutellata* is commonly called the eyelash cup fungus. *Morchella esculenta* is a common edible morel. *Helvella* is sometimes known as a saddle fungus since the fruiting body is thought by some to resemble a saddle.

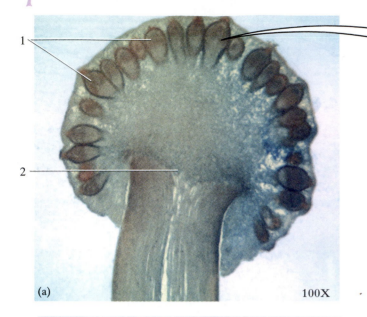

(a)   100X

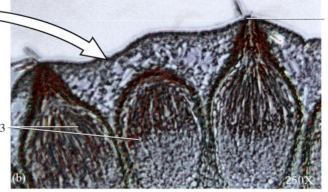

(b)   250X

Figure 5.15 The ascomycete, *Claviceps purpurea*. (a) Longitudinal section through stoma showing ascocarps (ascoma). (b) Enlargement of three perithecia. This fungus causes serious plant diseases and is toxic to humans.

1. Perithecia
2. Stroma within multiple perithecia
3. Perithecia containing asci
4. Ostiole

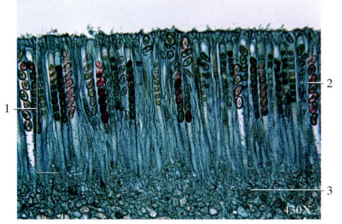

430X

Figure 5.16 A section through the hymenial layer of the apothecium of *Peziza* sp., showing asci with ascospores.

1. Hymenial layer
2. Ascus with ascospores
3. Ascocarp (ascoma) mycelium

3X

Figure 5.17 A section through an ascocarp (ascoma) of the morel, *Morchella* sp.. True morels are prized for their excellent flavor.

1. Convoluted fruiting body
2. Hollow "stalk"
3. Hymenium

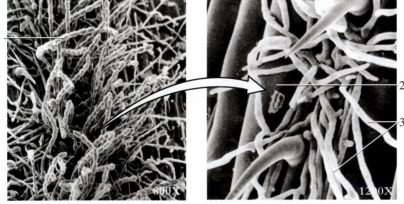

800X   1200X

Figure 5.18 A scanning electron micrographs of the powdery mildew, *Erysiphe graminis*, on the surface of wheat. As the mycelium develops, it produces spores (conidia) that give a powdery appearance to the wheat.

1. Conidia
2. Wheat host
3. Hyphae of the fungus

1800X

Photograph courtesy of: James V. Allen

Figure 5.19 A scanning electron micrograph of a germinating spore (conidium) of the powdery mildew, *Erysiphe graminis*. The spore develops into a mycelium that penetrates the epidermis and then spreads over the host plant, producing a powdery appearance.

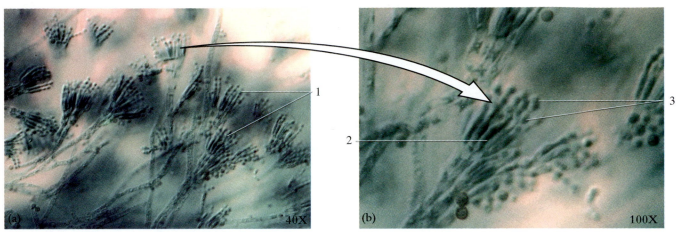

Figure 5.20 The fungus *Penicillium* sp. causes economic damage as a mold but is also the source of important antibiotics. (a) A colony of *Penicillium* sp., and (b) a close-up of a conidiophore with chains of asexual spores (conidia) at the end.

1. Conidia   2. Conidiophore   3. Conidia

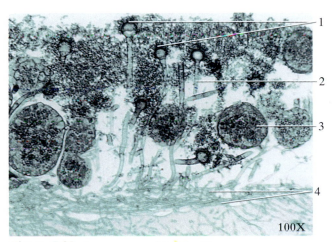

Figure 5.21 A common mold, *Aspergillus* sp.
1. Conidia (spores)   3. Cleistothecium
2. Conidiophore   4. Hyphae

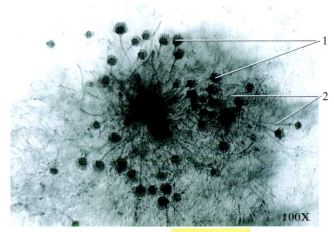

Figure 5.22 A common mold, *Aspergillus* sp.
1. Conidia   2. Conidiophores

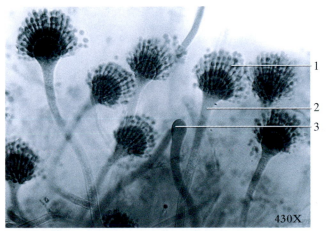

Figure 5.23 A close-up of sporangia of the mold, *Aspergillus* sp. The conidia, or spores, of this genus are produced in a characteristic radiate pattern.
1. Conidia (spores)   3. Developing conidiophore
2. Conidiophore

Photograph courtesy of: James V. Allen

Figure 5.24 An electron micrograph of an *Aspergillus* sp. spore. Note the rodlet pattern on the spore wall.

Phylum Basidiomycota - mushrooms, toadstools, rusts, and smuts

Pleurotus sp.

Hericium sp.

Coriolus sp.

Astreus sp.

Coprinus sp.

Amanita sp.

Chantarella sp.

Amanita sp.

Nidularia sp.

Boletus sp.

Figure 5.25 Some representative basidocarps (basidiomas or fruiting bodies) of basidiomycetes.

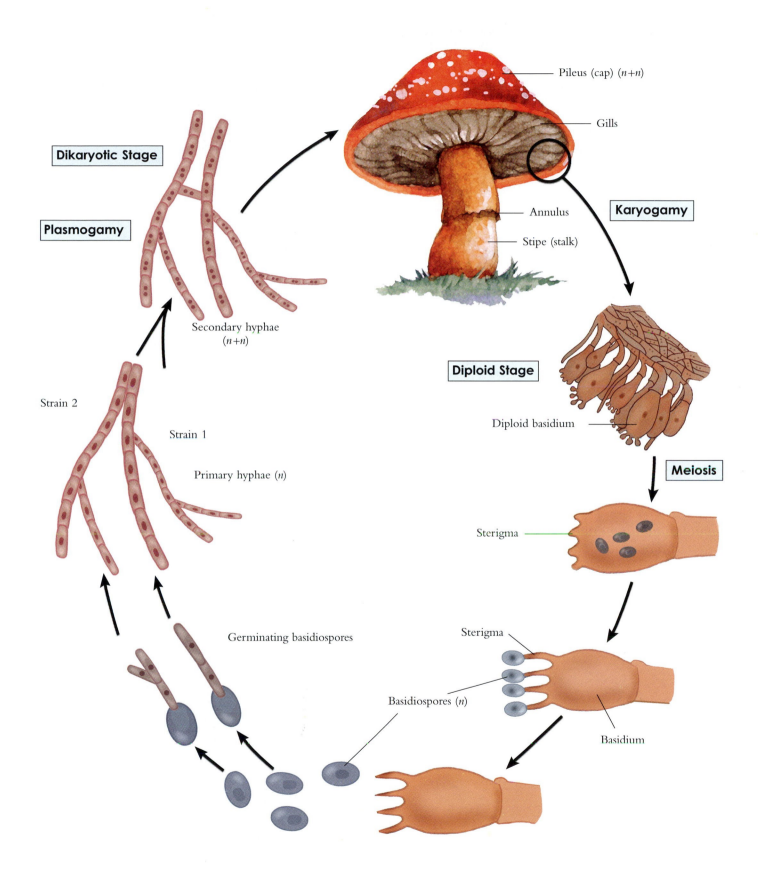

Pileus (cap) (*n+n*)

Gills

Annulus

Stipe (stalk)

Dikaryotic Stage

Plasmogamy

Karyogamy

Secondary hyphae
(*n+n*)

Diploid Stage

Diploid basidium

Strain 2

Meiosis

Strain 1

Sterigma

Primary hyphae (*n*)

Germinating basidiospores

Sterigma

Basidiospores (*n*)

Basidium

Figure 5.26 The life cycle of a "typical" basidiomycete (mushroom).

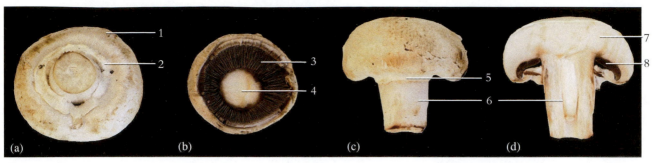

Figure 5.27 Structure of a mushroom. (a) An inferior view with the annulus intact, (b) an inferior view with the annulus removed to show the gills, (c) a lateral view, and (d) a longitudinal section.

1. Pileus (cap)   4. Stipe (stalk)   7. Pileus (cap)
2. Veil   5. Annulus   8. Gills
3. Gills   6. Stipe (stalk)

Figure 5.28 Basidiomycete puffballs growing on a decaying log.

Figure 5.29 The wood fungus, *Stropharia semiglobata*. Growing on decaying wood and other organic matter, basidiomycetes are important decomposers in forest communities.

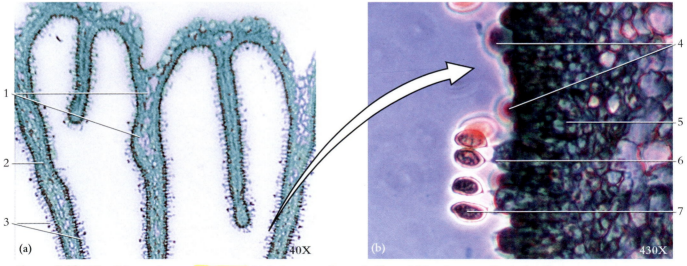

Figure 5.30 Gills of the mushroom *Coprinus* sp. (a) A close-up of several gills, and (b) a close-up of a single gill.

1. Hyphae comprising the gills   4. Immature basidia   7. Basidiospore
2. Gill   5. Gill (composed of hyphae)
3. Basidiospores   6. Sterigma

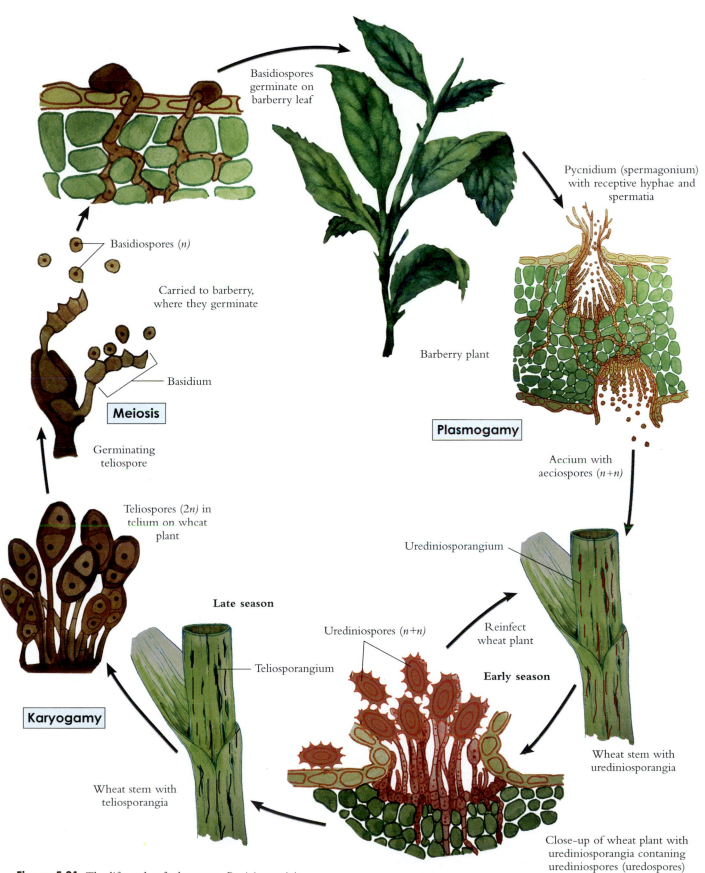

Basidiospores
germinate on
barberry leaf

Basidiospores (*n*)

Carried to barberry,
where they germinate

Basidium

Meiosis

Germinating
teliospore

Teliospores (2*n*) in
telium on wheat
plant

Karyogamy

Wheat stem with
teliosporangia

Late season

Teliosporangium

Pycnidium (spermagonium)
with receptive hyphae and
spermatia

Barberry plant

Plasmogamy

Aecium with
aeciospores (*n+n*)

Urediniosporangium

Urediniospores (*n+n*)

Reinfect
wheat plant

Early season

Wheat stem with
urediniosporangia

Close-up of wheat plant with
urediniosporangia contaning
urediniospores (uredospores)

Figure 5.31 The life cycle of wheat rust, *Puccinia graminis*.

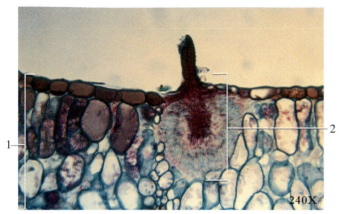

Figure 5.32 The wheat rust, *Puccinia graminis,* pycnidium on barberry leaf.
1. Barberry leaf
2. Pycnidium

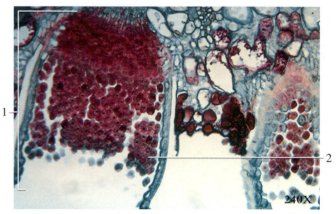

Figure 5.33 A *Puccinia graminis,* aecium on barberry leaf.
1. Aecium
2. Aeciospores

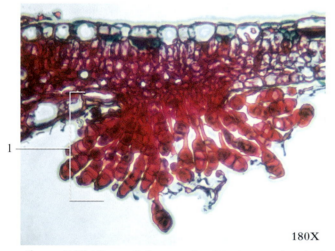

Figure 5.34 The urediniosporangia of *Puccinia* on wheat leaf.
1. Urediniosporangia

Figure 5.35 Black stem wheat rust, *Puccinia graminis,* on the lower surface of barberry leaves.
1. Clusters of aecia

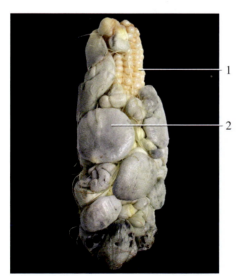

Figure 5.36 An ear of corn, *Zea mays,* infected by the smut *Ustilago maydis,* which is destroying the fruit (ear).
1. Corn ear
2. Fungus

Figure 5.37 Smut-infected brome grass. The grains have been destroyed by the smut fungus.

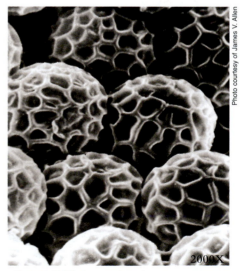

Photo courtesy of James V. Allen

Figure 5.38 A scanning electron micrograph of teliospores of a wheat smut fungus.

Lichens (symbiotic associations of fungi and algae)

Figure 5.39 Lichens are often separated informally on the basis of their form. (a) Crustose lichen, (b) foliose lichen, and (c) fruticose lichen.

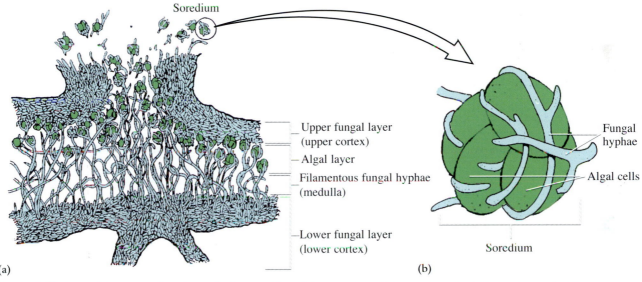

Soredium

Upper fungal layer
(upper cortex)

Algal layer

Filamentous fungal hyphae
(medulla)

Lower fungal layer
(lower cortex)

Fungal hyphae

Algal cells

Soredium

(a)   (b)

Figure 5.40 Many lichens reproduce by producing soredia, which are small bodies containing both algal and fungal cells. (a) Lichen thallus, and (b) soredium.

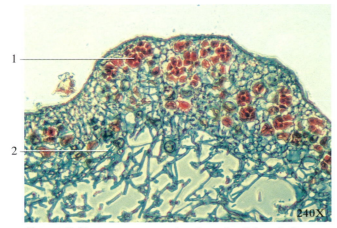 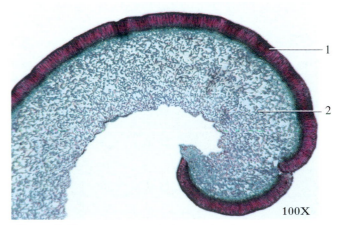

240X

100X

Figure 5.41 A transverse section through a lichen thallus.
  1. Algal cells   2. Fungal hyphae

Figure 5.42 An ascomycete lichen thallus demonstrating a surface layer of asci.
  1. Asci   2. Loose fungal filaments

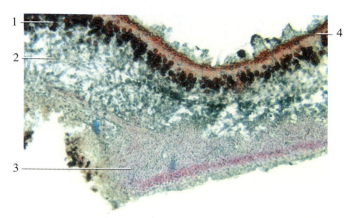

125X

Figure 5.43 A transverse section through a lichen thallus.
1. Algal cells
2. Medulla
3. Lower cortex
4. Fungal layer (upper cortex)

Figure 5.44 The foliose lichen *Xanthoria* sp. growing on the bark of a tree.
1. Lichen
2. Bark

Figure 5.45 The crustose lichen *Lecanora* sp. growing on sandstone in an arid southern Utah environment.

Figure 5.46 The foliose lichen *Hypogymnia* sp. growing on a pine branch in the Northwest.

Figure 5.47 The fruticose lichen *Alectoria* sp. growing in northern California.

Figure 5.48 The foliose and fruticose lichens in the Pacific Northwest.
1. Foliose lichen *Hypogymnia* sp.
2. Foliose lichen *Evernia* sp.
3. Fruticose lichen *Usnea* sp.

Plants are photosynthetic, multicellular eukaryotes. *Cellulose* in their cell walls provides protection and rigidity, while the *pores* or *stomata,* and *cuticle* of stems and leaves regulate gas exchange. Mitosis and meiosis are characteristic of all plants. Jacketed sex organs, called *gametangia*, protect the gametes and embryos from desiccation. All land plants have heteromorphic alternation of generations with distinctive haploid *gametophyte* and diploid *sporophyte* forms. Photosynthetic cells within plants contain *chloroplasts* with the pigments chlorophyll *a*, chlorophyll *b*, and a variety of carotenoids. Carbohydrates are produced by plants and stored in the form of starch.

Reproduction in seed plants is well adapted to a land existence. The conifers produce their seeds in protective *cones*, and the angiosperms produce their seeds in protective *fruits*. In the life cycle of a conifer, such as a pine, the mature *sporophyte* (tree) has female cones that produce *megaspores* that develop into the female gametophyte generation, and male cones that produce *microspores* that develop into the male gametophyte generation (mature pollen grains).

Following fertilization, immature sporophyte generations are present in seeds located on the female cones. The female cone opens and the *seeds* (pine nuts) disperse to the ground and germinate if the conditions are right. Reproduction in angiosperms is similar to reproduction in gymnosperms except that the angiosperm pollen and ovules are produced in flowers rather than in cones and a fruit is formed.

Table 6.1 Some Representatives of Plantae

Phyla (Division) and Representative Kinds	Characteristics
Bryophytes — liverworts, hornworts, and mosses	Lack vascular tissue; rhizoids; homosporous (bisexual gametophyte)
VASCULAR PLANTS **Lycopodiophyta** — clubmosses, spikemosses, and quillworts	Sporangia borne on sporophylls; homosporous or heterosporous (unisexual gametophyte); many are epiphytes
Pteridophyta Psilotopsida — whisk ferns	True roots and leaves are absent, but vascular tissue present; rhizome and rhizoids present
Pteridophyta Equisetopsida — horsetails	Epidermis embedded with silica; tips of stems bear cone-like structures containing sporangia; most homosporous
Pteridophyta Polypodiopsida — ferns	Fronds as leaves; underground roots coming off rhizomes; most homosporous
SEED PLANTS **Cycadophyta** — cycads	Heterosporous, pollen and seed cones borne of different plants; palm-like leaves
Ginkgophyta — ginkgo	Heterosporous; seed-producing; deciduous, fan-shaped leaves
Pinophyta (=Coniferophyta) — conifers	Heterosporous; pollen and seed cones same plant; needle-like or scale-like leaves
ANGIOSPERMS **Magnoliophyta (=Anthophyta)** — flowering plants	Heterosporous; flowering plants that produce their seeds enclosed in fruit; most are free-living, some are saprophytic or parasitic

Table 6.2 Some Representative Bryophytes

Marchantiophyta (=Hepatophyta) — liverworts	Flat or leafy gametophytes; single-celled rhizoids; simple sporophytes and elaters present; stomata and columella absent
Anthocerophyta — hornworts	Flat, lobed gametophytes; more complex sporophytes with stomata; pseudoelaters and columella present
Bryophyta — mosses	Leafy gametophytes, multicellular rhizoids; sporophytes with stomata, columella, peristome teeth and/or operculum present

Marchantiophyta (=Hepatophyta) - liverworts

Lepidozia   *Blepharostoma*   *Plectocolea*

Figure 6.1 An illustration of three genera of leafy liverworts, showing the gametophyte with an attached sporophyte. The perianth contains the archegonium and the lower portion of the developing sporophyte (yellowish).

Calopegia sp.   *Conocephalum* sp.   *Bazzania* sp.

Porella sp.   *Riccia* sp.   *Scapania* sp.

Figure 6.2 Some examples of liverworts (scale in mm).

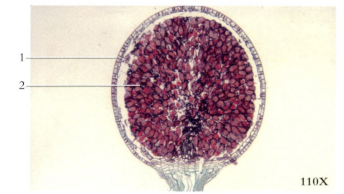

110X

Figure 6.3 A sporophyte (capsule) of the leafy liverwort, *Pelia* sp.

1. Capsule   2. Sporogenous tissue

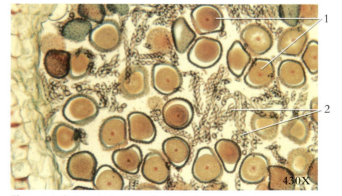

430X

Figure 6.4 A capsule from the leafy liverwort, *Pelia* sp., in longitudinal view.

1. Spores   2. Elaters

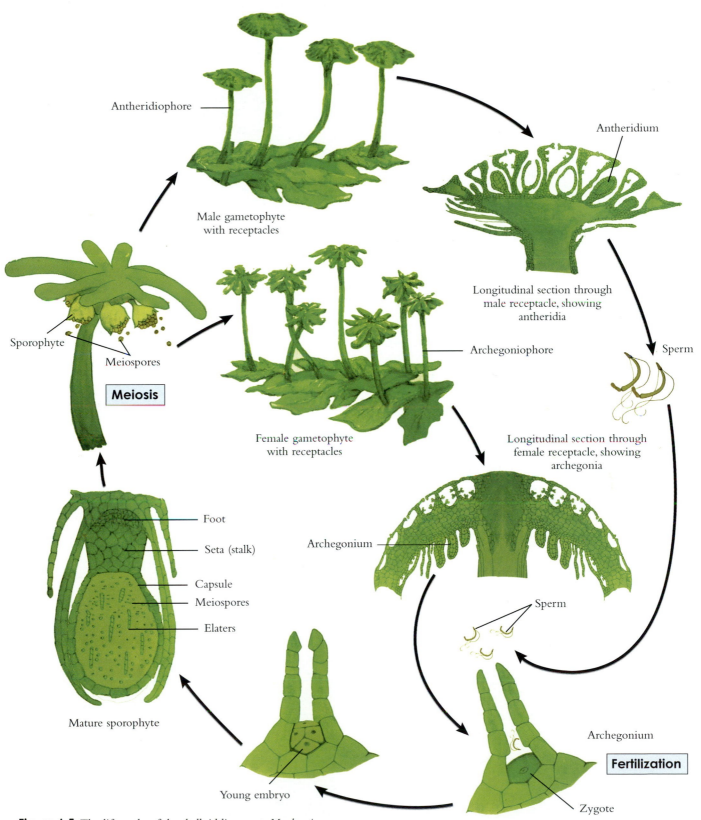

Antheridiophore

Male gametophyte
with receptacles

Antheridium

Longitudinal section through
male receptacle, showing
antheridia

Sporophyte

Meiospores

Meiosis

Archegoniophore

Sperm

Female gametophyte
with receptacles

Longitudinal section through
female receptacle, showing
archegonia

Foot

Seta (stalk)

Archegonium

Capsule

Meiospores

Elaters

Sperm

Mature sporophyte

Archegonium

Young embryo

Zygote

Fertilization

Figure 6.5 The life cycle of the thalloid liverwort, *Marchantia* sp.

Figure 6.6 A detail of *Marchantia* sp. with prominent male antheridial receptacles.
1. Antheridial receptacles
2. Gametophyte thallus

Figure 6.7 The liverwort *Marchantia* sp., showing archegonial receptacles.
1. Archegonial receptacles

Figure 6.8 A detail of *Marchantia* sp. gametophyte plants with prominent gemmae cupules.
1. Gemmae cup with gemmae

Figure 6.9 A transverse section through a gemma cupule of *Marchantia* sp.
1. Gemmae cupule   2. Gemmae

Figure 6.10 The liverwort *Marchantia* sp., showing rhizoids.
1. Rhizoids

Photo courtesy of James V. Allen

1000X

Figure 6.11 A scanning electron micrograph of the thallus of *Marchantia* sp.
1. Air pore

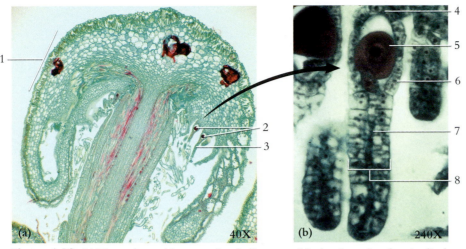

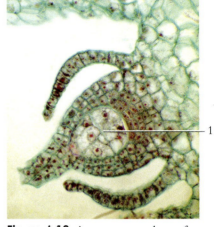

Figure 6.12 (a) The archegonial receptacle of a liverwort, *Marchantia* sp., in a longitudinal section. (b) Archegonium with egg.

1. Archegonial receptacle
2. Eggs
3. Neck of archegonium
4. Base of archegonium
5. Egg
6. Venter of archegonium
7. Neck canal
8. Neck of archegonium

Figure 6.13 A young sporophyte of *Marchantia* sp.

1. Young embryo

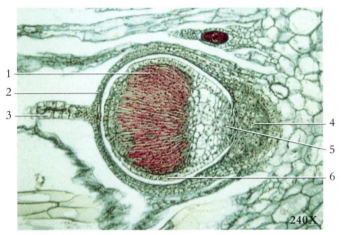

Figure 6.14 A young sporophyte of *Marchantia* sp., in longitudinal section.

1. Sporogenous tissue (*2n*)
2. Enlarged archegonium (calyptra)
3. Neck of archegonium
4. Foot
5. Seta (stalk)
6. Capsule

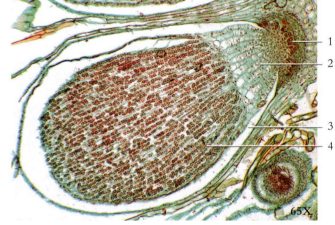

Figure 6.15 Immature and mature sporophytes.

1. Foot
2. Seta (stalk)
3. Sporangium (capsule)
4. Spores (*n*) and elaters (*2n*)

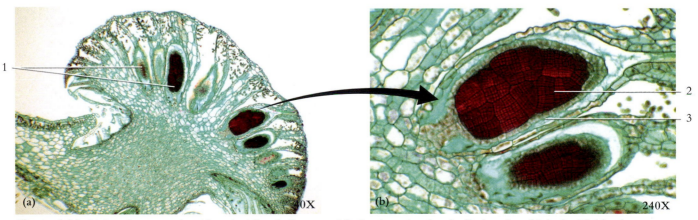

Figure 6.16 (a) A male receptacle with antheridia of a liverwort, *Marchantia* sp., in a longitudinal section. (b) Antheridial head showing a developing antheridium.

1. Antheridia
2. Spermatogenous tissue
3. Antheridium

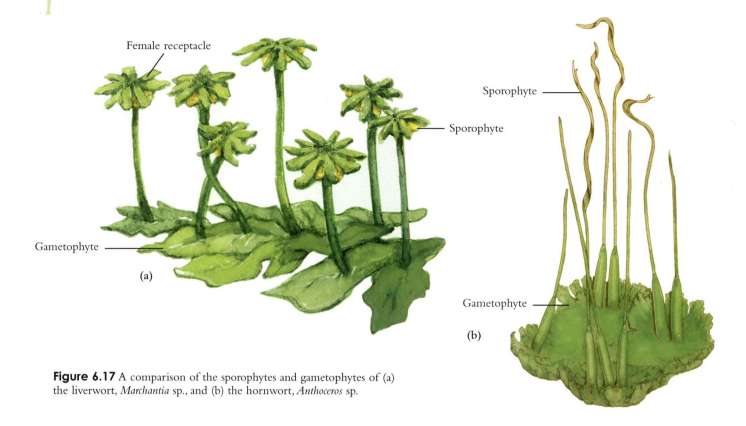

Female receptacle

Sporophyte

Sporophyte

Gametophyte

Gametophyte

(a)

(b)

Figure 6.17 A comparison of the sporophytes and gametophytes of (a) the liverwort, *Marchantia* sp., and (b) the hornwort, *Anthoceros* sp.

Anthocerophyta - hornworts

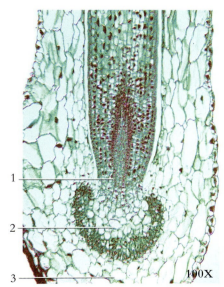

100X

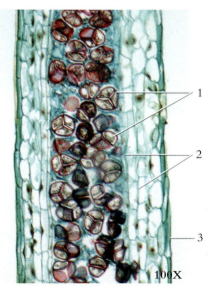

100X

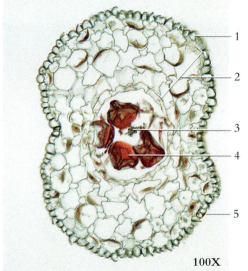

100X

Figure 6.18 A longitudinal section of a portion of the sporophyte of the hornwort, *Anthoceros* sp.
1. Meristematic region of sporophyte
2. Foot
3. Gametophyte

Figure 6.19 A longitudinal section of the sporangium of a sporophyte from the hornwort, *Anthoceros* sp.
1. Spores
2. Elater-like structures (pseudoelaters)
3. Capsule

Figure 6.20 A transverse section through the capsule of a sporophyte of the hornwort, *Anthoceros* sp.
1. Epidermis
2. Photosynthetic tissue
3. Columella
4. Tetrad of spores
5. Pore (stomate)

Bryophyta - mosses

Figure 6.21 A *Sphagnum* bog in the high Rocky Mountains. This lake has nearly been filled in with dense growths of *Sphagnum* sp.

Figure 6.22 A detail of *Sphagnum* bog showing gametophyte plants.

Figure 6.23 A detail of gametophyte plants of peat moss, *Sphagnum* sp. (scale in mm).

Figure 6.24 A gametophyte plant of peat moss, *Sphagnum* sp., showing attached sporophytes (scale in mm).

1. Sporophyte   3. Gametophyte
2. Pseudopodium

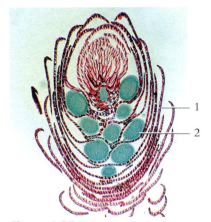

Figure 6.25 A longitudinal section of *Sphagnum* sp. gametophyte showing antheridia.

1. "leaf"   2. Antheridium

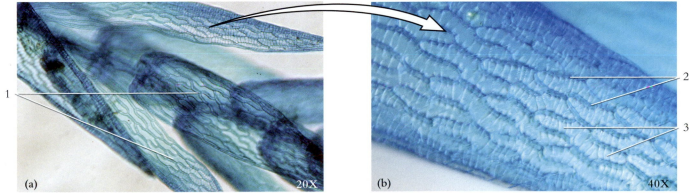

Figure 6.26 (a) A gametophyte of peat moss, *Sphagnum* sp. (b) A magnified view of a "leaf", showing the dead cell chambers that aid in water storage.

1. "Leaves"   2. Photosynthetic cells   3. Dead cells

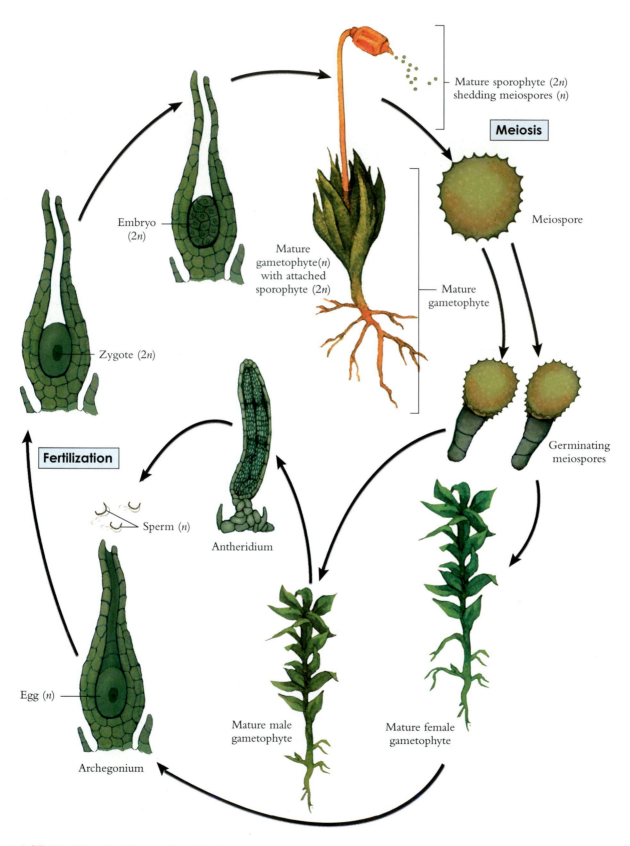

Mature sporophyte (2n) shedding meiospores (n)

Meiosis

Meiospore

Embryo (2n)

Mature gametophyte(n) with attached sporophyte (2n)

Mature gametophyte

Zygote (2n)

Germinating meiospores

Fertilization

Sperm (n)

Antheridium

Egg (n)

Archegonium

Mature male gametophyte

Mature female gametophyte

Figure 6.27 The life cycle of a moss (Bryophyta).

Figure 6.28 A habitat shot of a moss growing in a wooded environment.
1. Moss
2. Vascular plants

Figure 6.29 A moss-covered sand stone. Under dry conditions, mosses may become dormant and lose their intense green.
1. Stone   2. Moss

Figure 6.30 Four common mosses often used in course work, (a) *Polytrichum* sp., (b) *Mnium* sp., (c) *Hypnum* sp. and (d) *Dicranum* sp.

Figure 6.31 Gametophyte plants with sporophyte plant attached.
1. Calyptera
2. Capsule of sporophyte (covered by calyptera)
3. Stalk (seta)
4. Gametophyte

Figure 6.32 A sporophyte plant and capsule.
1. Operculum
2. Capsule of sporophyte (with calyptera absent)
3. Stalk (seta)

Figure 6.33 The protonemata and bulbils of a moss. The bulbils will grow to become a new gametophyte plant.
1. Protonema   2. Bulbil

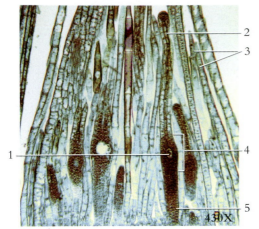

Figure 6.34 A longitudinal section of the archegonial head of the moss *Mnium* sp. The paraphyses are non-reproductive filaments that support the archegonia.

1. Egg   3. Paraphyses   5. Stalk
2. Neck   4. Venter

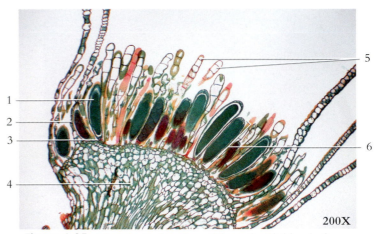

Figure 6.35 A longitudinal section of the antheridial head of the moss *Mnium* sp.

1. Spermatogenous tissue   4. Male gametophyte (*n*)
2. Sterile jacket layer   5. Paraphyses (sterile filaments)
3. Stalk   6. Antheridium (*n*)

Figure 6.36 A close-up of *Mnium* sp.

1. Antheridium (*n*)   3. Stalk
2. Spermatogenous tissue   4. Paraphyses

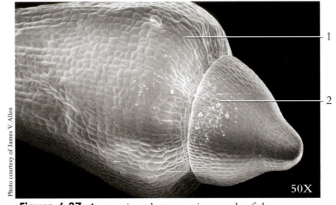

Photo courtesy of James V. Allen

Figure 6.37 A scanning electron micrograph of the sporophyte capsule of the moss *Mnium* sp.

1. Capsule   2. Operculum

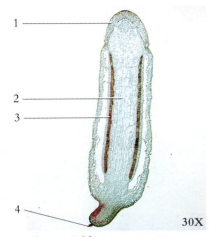

Figure 6.38 A capsule of the moss *Mnium* sp.

1. Operculum   3. Spores
2. Columella   4. Seta

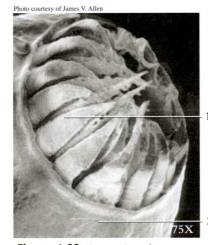

Photo courtesy of James V. Allen

Figure 6.39 A scanning electron micrograph of the peristome of the moss *Mnium* sp. The operculum is absent in the specimen.

1. Peristome   2. Capsule

Photo courtesy of James V. Allen

Figure 6.40 A scanning electron micrograph of the peristome of the moss *Mnium* sp.

1. Outer teeth   3. Inner teeth
   of peristome   of peristome
2. Capsule

Lycopodiophyta (Lycophyta) - clubmosses, spikemosses, and quillworts

Strobilus

Longitudinal section through strobilus

Sporangium

Sporophyll

Mature sporophyte (2n)

Meiosis

Older embryo (2n)

Young embryo (2n)

Meiospores (n)

Zygote (2n)

Sporangium

Sperm (n)

Meiospores (n)

Fertilization

Antheridia

Egg (n)

Archegonium

Mature gametophyte (n) (often underground)

Longitudinal section through gametophyte with gametangia

Figure 6.41 The life cycle of the homosporous clubmoss, *Lycopodium* sp.

Figure 6.42 A specimen of a lycopod, *Lycopodium clavatum*, (a) plant and (b) strobilus. *Lycopodium* occurs from the arctic to the tropics (scale in mm).
  1. Strobilus   2. Stem

Figure 6.43 An enlargement of a specimen of *Lycopodium* sp., showing branch tip with sporangia on the upper surface of sporophylls (scale in mm).
  1. Sporangia
  2. Sporophylls (leaves with attached sporangia)

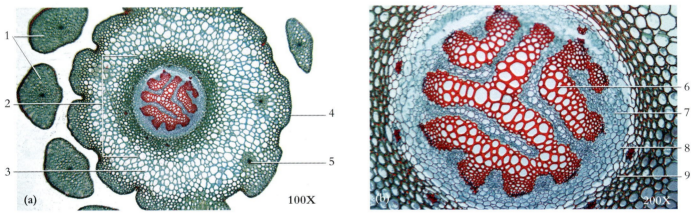

Figure 6.44 (a) A transverse view of an aerial stem of the clubmoss, *Lycopodium* sp. (b) A magnified view of the stele.
  1. Leaves (microphylls)   4. Epidermis   5. Leaf trace   7. Phloem   9. Endodermis
  2. Stele   3. Cortex   6. Xylem   8. Pericycle

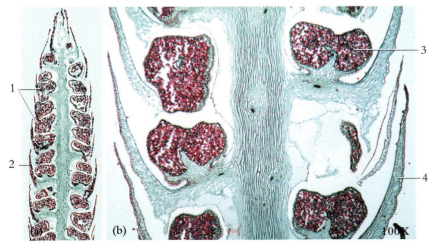

Figure 6.45 (a) A longitudinal section of the strobilus (cone) of the clubmoss *Lycopodium* sp., and (b) a magnified view of the strobilus showing sporangia.
  1. Sporangia   3. Sporangium
  2. Sporophyll   4. Sporophyll

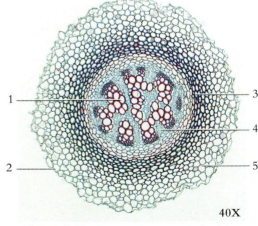

Figure 6.46 A transverse section of a rhizome of *Lycopodium* sp. The rhizome of *Lycopodium* is similar to an aerial stem, but it lacks the microphylls.
  1. Xylem   3. Endodermis   5. Cortex
  2. Epidermis   4. Phloem

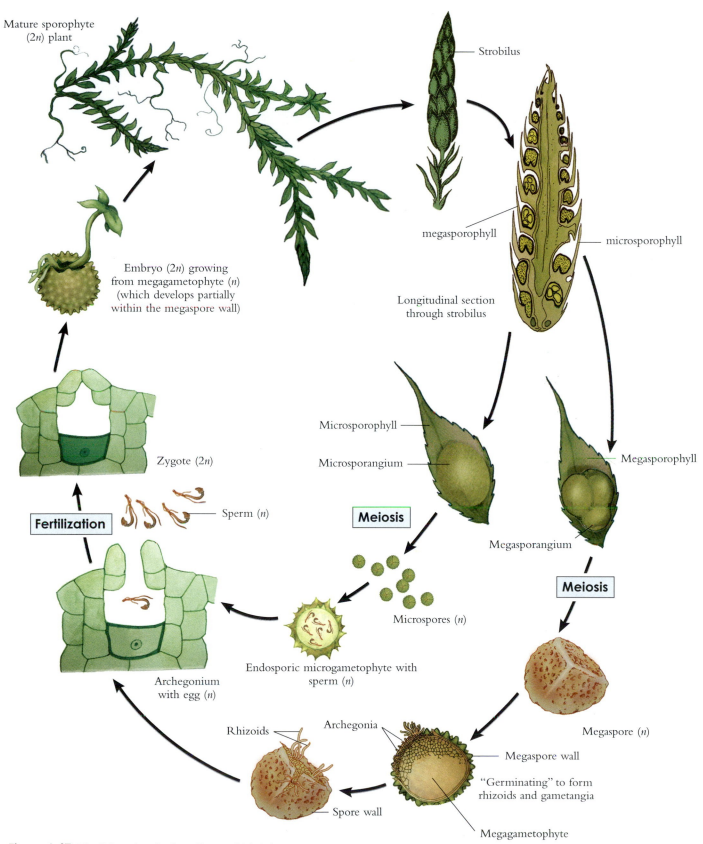

Mature sporophyte
(2n) plant

Strobilus

megasporophyll

microsporophyll

Longitudinal section
through strobilus

Embryo (2n) growing
from megagametophyte (n)
(which develops partially
within the megaspore wall)

Microsporophyll

Microsporangium

Megasporophyll

Zygote (2n)

Megasporangium

Meiosis

Sperm (n)

Fertilization

Meiosis

Microspores (n)

Endosporic microgametophyte with
sperm (n)

Archegonium
with egg (n)

Megaspore (n)

Rhizoids

Archegonia

Megaspore wall

"Germinating" to form
rhizoids and gametangia

Spore wall

Megagametophyte

Figure 6.47 The life cycle of *Selaginella* sp., which is heterosporous.

Figure 6.48 The spikemoss, *Selaginella kraussiana* (a) growth habit and (b) strobili (cones).
1. Strobili (cones)   2. Sporaphyll with sporangium

Figure 6.49 The spikemoss, *Selaginella pulcherrima*.

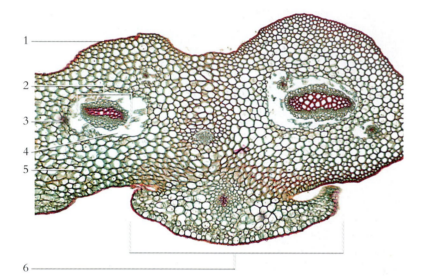

200X

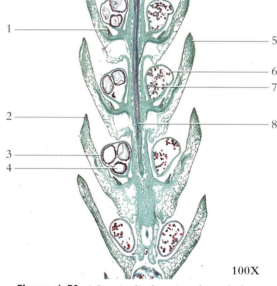

100X

Figure 6.50 A transverse section through stem of *Selaginella* sp. immediately above dichotomous branching.
1. Epidermis
2. Protostele (surrounded by endodermis)
3. Root trace
4. Air cavity
5. Cortex
6. Leaf base

Figure 6.51 A longitudinal section through the strobilus of *Selaginella* sp.
1. Ligule
2. Megasporophyll
3. Megasporangium
4. Megaspore
5. Microsporophyll
6. Microsporangium
7. Microspore
8. Cone axis

20X

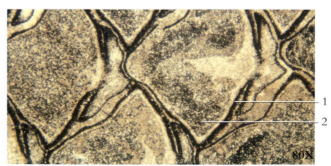

80X

Figure 6.52 A longitudinal view of the surface of the fossil lycophyte *Lepidodendron* sp., a common lycopod from perhaps 300 million years ago.

Figure 6.53 A longitudinal section through a fossil strobilus of the lycophyte *Lepidostrobus* sp., from approximately 300 million years ago.
1. Sporangium   2. Sporogenous tissue

Pteridophyta: Psilotopsida - whisk ferns

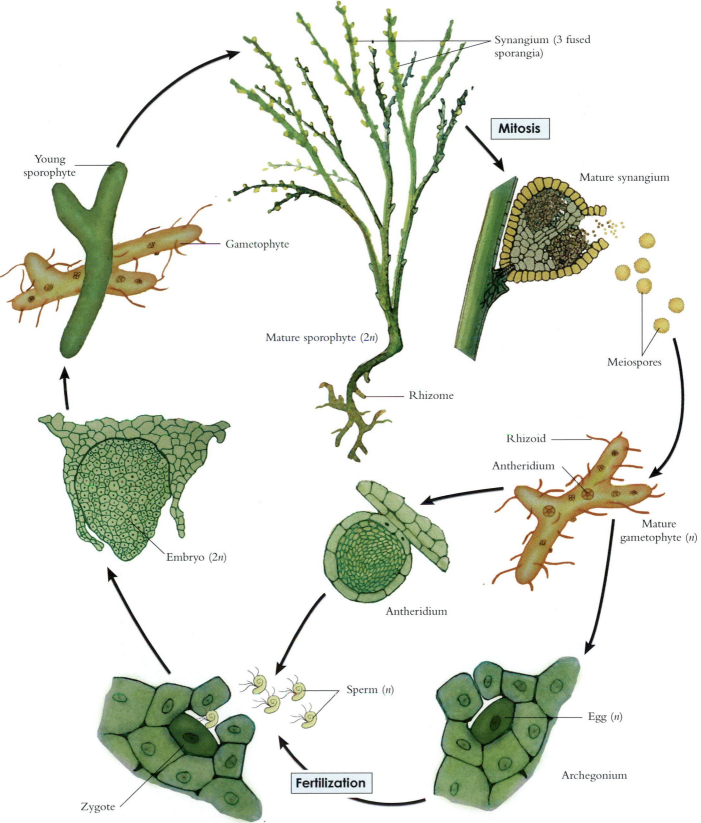

Synangium (3 fused sporangia)

Mitosis

Mature synangium

Young sporophyte

Gametophyte

Mature sporophyte (2n)

Meiospores

Rhizome

Rhizoid

Antheridium

Mature gametophyte (n)

Embryo (2n)

Antheridium

Sperm (n)

Egg (n)

Archegonium

Fertilization

Zygote

Figure 6.54 The life cycle of the whisk fern, *Psilotum* sp.

Figure 6.55 A *Tmesipteris* sp., growing as an epiphyte on a tree fern in Australia.

Figure 6.56 A whisk fern, *Psilotum nudum,* is a simple vascular plant lacking true leaves and roots.

Figure 6.57 The branches (axes) of *Psilotum nudum* (scale in mm) .
1. Aerial axis   2. Rhizome

Figure 6.58 A sporophyte of the whisk fern, *Psilotum nudum.* The axes of the sporophyte support sporangia (synangia), which produce spores (scale in mm).
1. Branch (axis)   2. Sporangia (synangia)

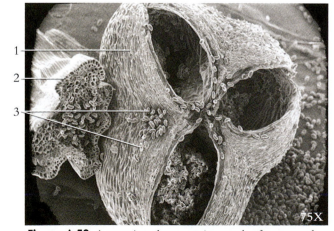

Figure 6.59 A scanning electron micrograph of a ruptured synangium (3 fused sporangia) of *Psilotum* sp., which is spilling spores.
1. Sporangium (often   2. Axis   3. Spores
   called synangia)

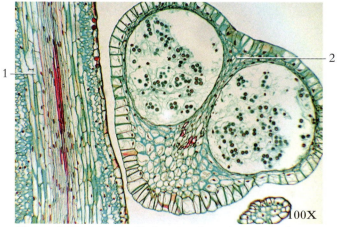

Figure 6.60 A longitudinal section through a stem and sporangium (synangium) of *Psilotum* sp.
1. Axis   2. Sporangia (synangium)

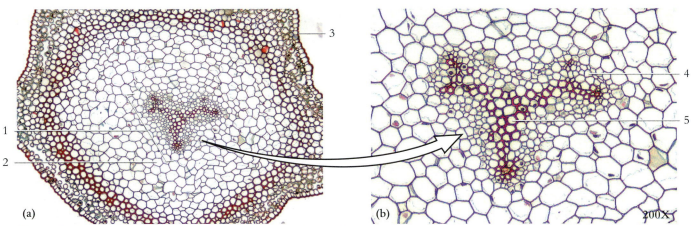

Figure 6.61 An aerial axis of the whisk fern, *Psilotum nudum.* (a) Transverse section and (b) a magnified view of the vascular cylinder (stele).

| 1. Stele | 2. Cortex | 3. Epidermis | 4. Phloem | 5. Xylem |

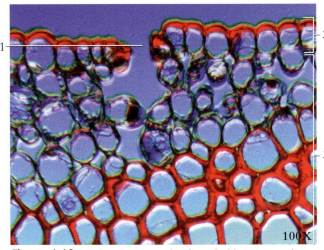

Figure 6.62 A photomicrograph of a scale-like outgrowth from the axis of the whisk fern *Psilotum nudum.*

| 1. Stoma | 2. Epidermis | 3. Ground tissue |

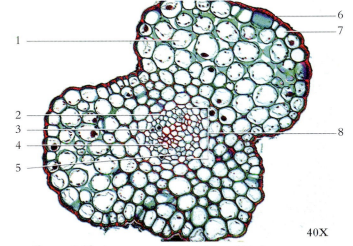

Figure 6.63 A young aerial axis of the whisk fern, *Tmesipteris* sp.

1. Cortex	4. Xylem	7. Epidermis
2. Endodermis	5. Phloem	8. Protostele
3. Pericycle	6. Cuticle	

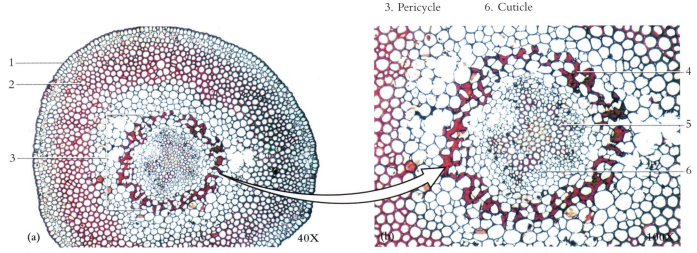

Figure 6.64 An older aerial axis of the whisk fern, *Tmesipteris* sp. The genus *Tmesipteris* is restricted to distribution in Australia, New Zealand, New Caledonia, and other South Pacific islands. (a) Axis arising from the aerial axis and (b) a magnified view of the stele.

| 1. Epidermis | 2. Cortex | 3. Stele | 4. Endodermis | 5. Xylem | 6. Phloem |

Pteridophyta: Equisetopsida (=Sphenopsida) - horsetails

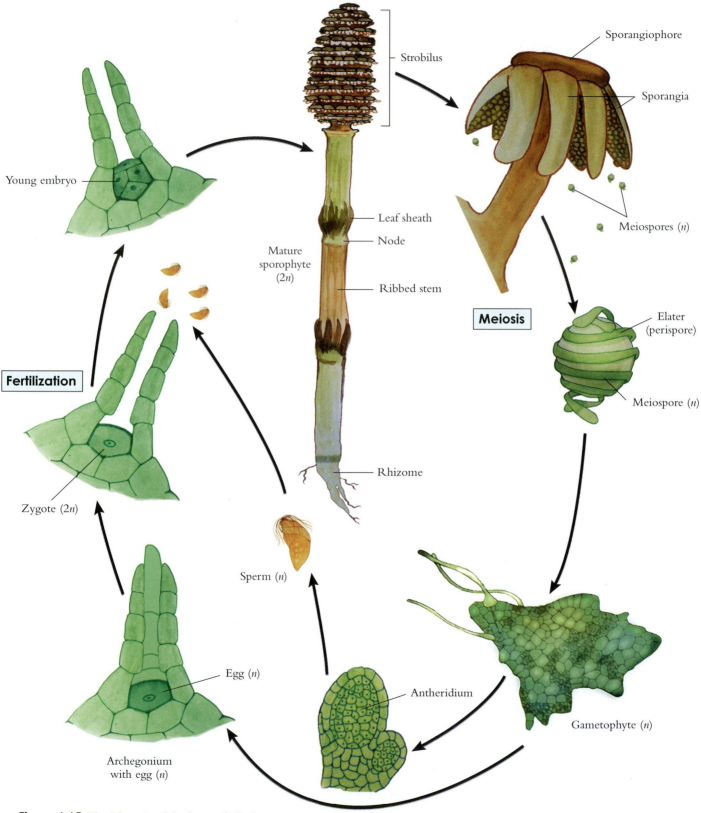

Sporangiophore

Sporangia

Strobilus

Young embryo

Meiospores (*n*)

Leaf sheath

Node

Mature
sporophyte
(2*n*)

Ribbed stem

Meiosis

Elater
(perispore)

Meiospore (*n*)

Fertilization

Zygote (2*n*)

Rhizome

Sperm (*n*)

Egg (*n*)

Antheridium

Gametophyte (*n*)

Archegonium
with egg (*n*)

Figure 6.65 The life cycle of the horsetail, *Equisetum* sp.

Figure 6.66 An *Equisetum telmateia* showing lateral branching.

Figure 6.67 A close-up of *Equisetum telmateia* showing lateral branches growing through leaf sheath.

Figure 6.68 The stems of *Equisetum* sp. without lateral branching and showing a prominent leaf sheath at the node.
1. Stem   2. Leaf sheath

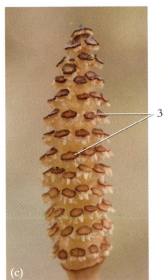

(a)   (b)   (c)   (d)

Figure 6.69 The horsetail, *Equisetum* sp. Numerous species of Equisetophyta were abundant throughout tropical regions during the Paleozoic Era, some 300 million years ago. Currently, Equisetophyta are represented by this single genus. The meadow horsetail, *Equisetum* sp., showing (a) an immature strobilus, (b) mature strobilus, shedding spores, (c) an open strobilus, and (d) a sporangiophore with its spores released.
1. Sporangiophores
2. Separated sporangiophores revealing sporangia
3. Sporangiophores after spores are shed
4. Open sporangia with spores shed

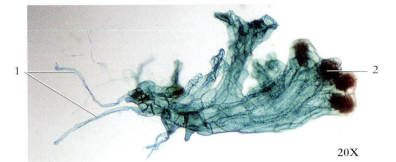

20X

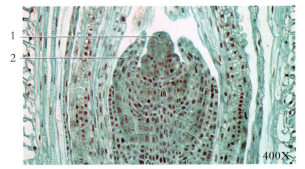

400X

Figure 6.70 A young gametophyte of *Equisetum* sp.
1. Rhizoids   2. Antheridium

Figure 6.71 A longitudinal section of *Equisetum* sp. shoot apex.
1. Apical cell   2. Leaf primordium

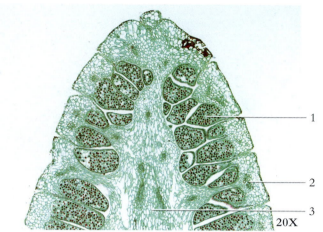

Figure 6.72 A longitudinal section of *Equisetum* sp. strobilus.
1. Sporangium   2. Sporangiophore   3. Strobilus axis

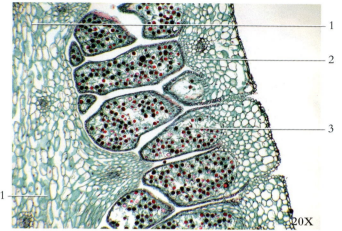

Figure 6.73 A longitudinal section through *Equisetum* sp. strobilus.
1. Axis of the strobilus   2. Sporangiophore   3. Sporangium

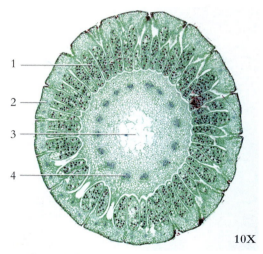

Figure 6.74 A transverse section of the strobilus of *Equisetum* sp.
1. Sporangium   3. Strobilus axis
2. Sporangiophore   4. Vascular bundle

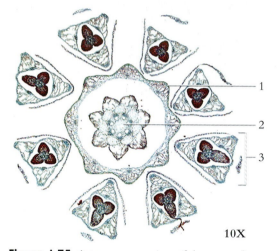

Figure 6.75 A transverse section of the stem of *Equisetum* sp. just above a node.
1. Leaf sheath   2. Main stem   3. Branch

Figure 6.76 The meiospores of *Equisetum* sp.
1. Perispore (elater)
2. Meiospore

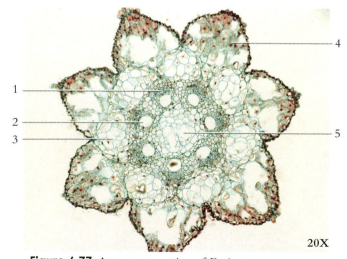

Figure 6.77 A transverse section of *Equisetum* sp. young stem.
1. Vascular tissue   3. Future air canal   5. Pith
2. Air canal   4. Cortex

Figure 6.78 A transverse section of *Equisetum* sp. older stem.
1. Air canals   3. Vascular tissue   5. Eustele
2. Endodermis   4. Stomate

Pteridophyta: Polypodiopsida - ferns

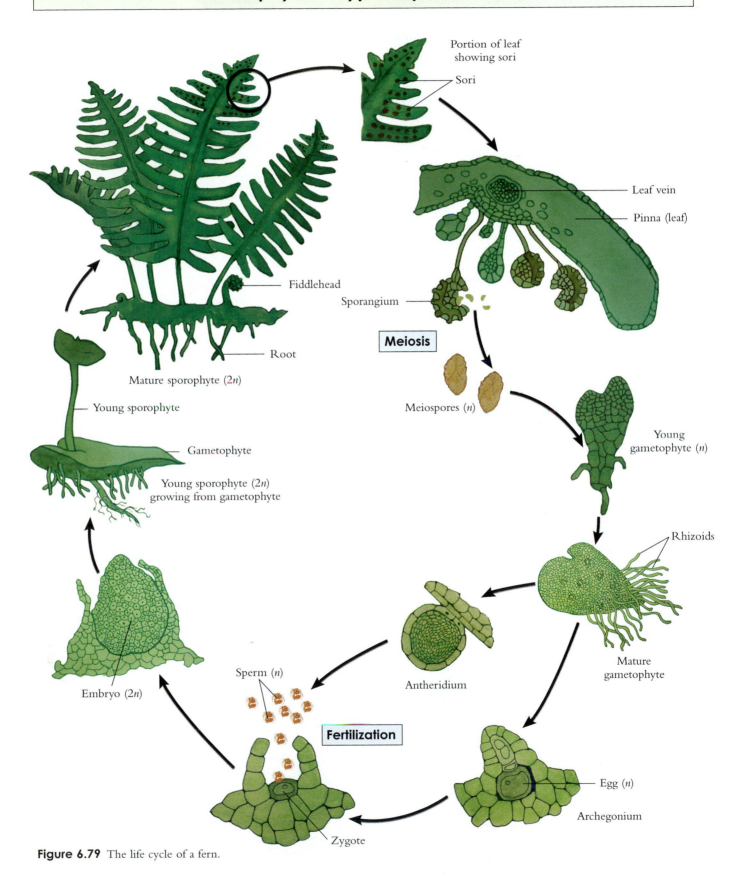

Portion of leaf
showing sori

Sori

Leaf vein

Pinna (leaf)

Sporangium

Meiosis

Meiospores (*n*)

Young
gametophyte (*n*)

Rhizoids

Fiddlehead

Root

Mature sporophyte (2*n*)

Young sporophyte

Gametophyte

Young sporophyte (2*n*)
growing from gametophyte

Mature
gametophyte

Antheridium

Sperm (*n*)

Embryo (2*n*)

Fertilization

Egg (*n*)

Archegonium

Zygote

Figure 6.79 The life cycle of a fern.

Figure 6.80 The water fern, *Azolla* sp., is a floating freshwater plant found throughout Europe and the United States. *Azolla* may become bright orange or red during fall.

Figure 6.81 A view of a new (a) compound and (b) simple fern leaf showing circinate vernation forming a fiddlehead.

Figure 6.82 The fronds of the staghorn fern, *Platycerium alcicorne.*

Figure 6.83 A pinnate leaf showing pinnate venation in the leaflets of a fern.
1. Leaf   2. Pinnae   3. Venation

Figure 6.84 A leaf of the fern *Phanerophlebia* sp., or holly fern.

Figure 6.85 A leaf of the fern *Phanerophlebia* sp., showing sori (groups of sporangia).
1. Pinna
2. Sori

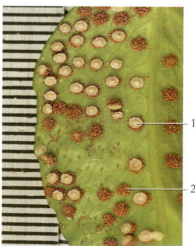

Figure 6.86 A close-up of the fern leaf of *Phanerophlebia* sp. (scale in mm).
1. Sorus with indusium
2. Sorus with indusium shed

Figure 6.87 The leaf of the fern *Polypodium virginianum*.

Figure 6.88 The leaf of the fern *Polypodium virginianum*, showing sori (groups of sporangia).
1. Pinna   2. Sori

Figure 6.89 A close-up of the fern pinna of *Polypodium virginianum* (scale in mm).
1. Sorus

Figure 6.90 The fern *Polypodium* sp. (a) Sori on the undersurface of the pinnae, and (b) a scanning electron micrograph of a sorus.
1. Pinna   2. Sori   3. Annulus   4. Sporangium

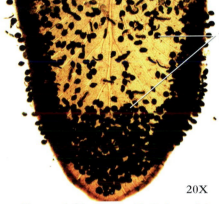

Figure 6.91 A magnified view of the fern pinna of *Pteridium* sp. showing numerous scattered sporangia.
1. Sporangia

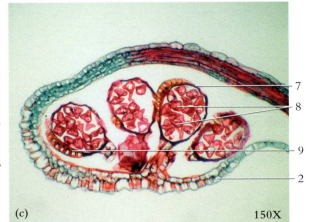

Figure 6.92 The maidenhair fern *Adiantum* sp. (a) Pinnae and sori. (b) Magnified view of the tip of a pinna folded under to form a false indusium that encloses the sorus. (c) Sorus with sporangia containing spores (scale in mm).
1. False indusium   3. Pinna   5. False indusium enclosing a sorus   7. Sporangium   9. Annulus
2. Sori   4. Sporangia with spores   6. Vascular tissue (veins) of the pinna   8. Spores

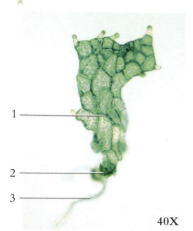

40X

Figure 6.93 A young fern gametophyte.
1. Gametophyte
2. Spore cell wall
3. Rhizoid

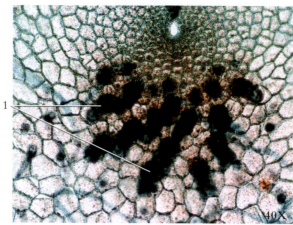

40X

Figure 6.94 A fern gametophyte with archegonia.
1. Archegonia

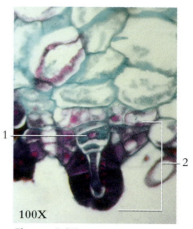

100X

Figure 6.95 A fern gametophyte showing archegonium.
1. Egg
2. Archegonium

100X

Figure 6.96 A fern gametophyte showing antheridia.
1. Gametophyte (prothallus)
2. Antheridium with sperm
3. Rhizoids

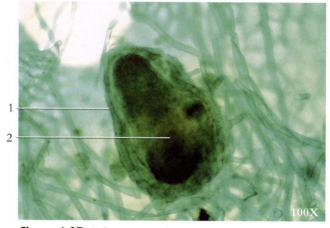

100X

Figure 6.97 A fern gametophyte with a young sporophyte attached.
1. Expanded archegonium
2. Young sporophyte

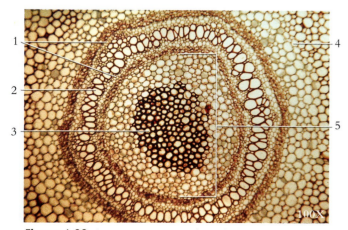

100X

Figure 6.98 A transverse section through the stem of a fern, *Dicksonia* sp. showing a siphonostele.
1. Phloem
2. Xylem
3. Sclerified pith
4. Cortex
5. Pith

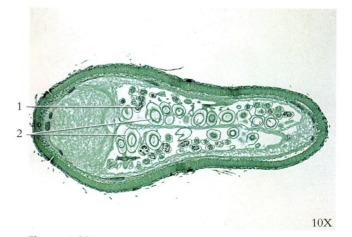

10X

Figure 6.99 A transverse section of a sporocarp of the water fern, *Marsilea* sp., which is one of the two living orders of heterosporous ferns.
1. Microsporangium with microspores
2. Megasporangia with megaspores

Cycadophyta - cycads

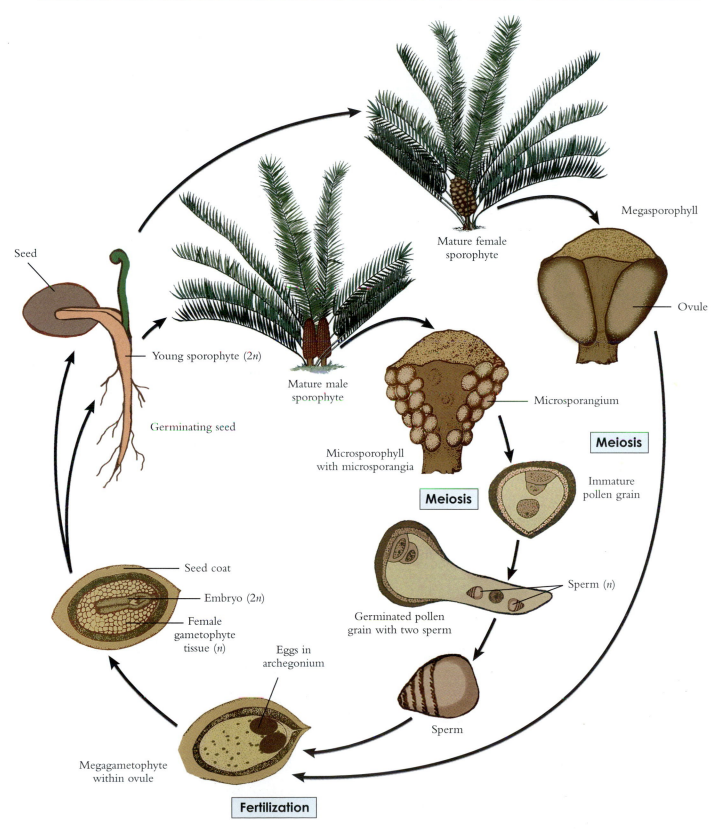

Seed

Young sporophyte (2n)

Germinating seed

Mature male
sporophyte

Mature female
sporophyte

Megasporophyll

Ovule

Microsporophyll
with microsporangia

Microsporangium

Meiosis

Immature
pollen grain

Meiosis

Sperm (n)

Germinated pollen
grain with two sperm

Sperm

Seed coat

Embryo (2n)

Female
gametophyte
tissue (n)

Eggs in
archegonium

Megagametophyte
within ovule

Fertilization

Figure 6.100 The life cycle of a cycad.

Figure 6.101 A *Cycas revoluta*. Cycads were abundant during the Mesozoic Era. Currently, there are 10 living genera, with about 100 species, that are found mainly in tropical and subtropical areas. The trunk of many cycads is densely covered with petioles of shed leaves.

Figure 6.102 A *Cycas revoluta* showing a female cone.
1. Cone

Figure 6.103 A *Cycas revoluta* showing a close-up view of a female cone with developing seeds.
1. Seeds   2. Megasporophyll

Figure 6.104 A *Cycas revoluta* showing a close-up view of a female cone during seed dispersal.
1. Seeds

Figure 6.105 A male cone of *Cycas revoluta*.
1. Cone

Figure 6.106 A male cone of *Cycas revoluta* after release of pollen.

Figure 6.107 A young plant of the cycad *Zamia pumila*. Found in Florida, this cycad is the only species native to the United States. The rootstocks and stems of this plant were an important source of food for some Native Americans.

Figure 6.108 A microsporangiate cones of the cycad *Zamia* sp.

Figure 6.109 The *Encephalartos villosus* is a non-threatened species of cycad native to southeastern Africa.

Figure 6.110 A maturing female cone of *Encephalartos villosus*.

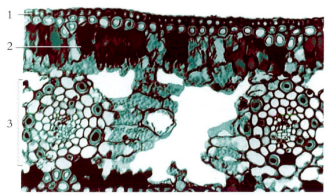

Figure 6.111 A transverse section of the leaf of the cycad *Zamia* sp.
1. Upper epidermis
2. Palisade mesophyll
3. Vascular bundle (vein)

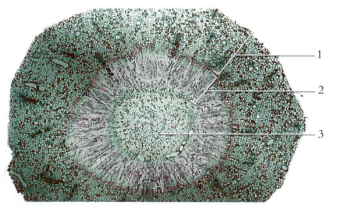

Figure 6.112 A transverse section of the stem of the cycad *Zamia* sp.
1. Cortex   3. Pith
2. Vascular tissue

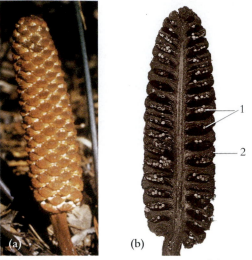

Figure 6.113 A microsporangiate cone of the cycad, *Zamia* sp. The cone on the right (b) is longitudinally sectioned.

1. Microsporangia   2. Microsporophyll

Figure 6.114 A microsporangiate cone of a cycad showing megasporangia on microsporophylls.

1. Microsporangium   2. Microsporophyll

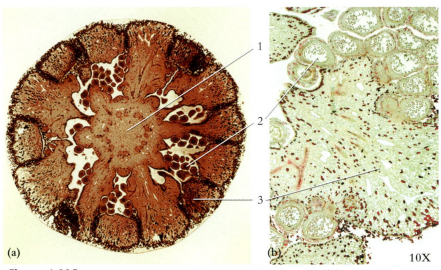

Figure 6.115 A transverse sections of a microsporangiate cone of the cycad *Zamia* sp. (a) A low magnification, and (b) a magnified view.

1. Cone axis   2. Microsporangia   3. Microsporophyll

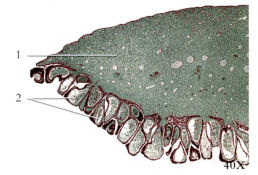

Figure 6.116 A longitudinal section of a microsporophyll of the cycad *Cycas* sp. Note that the microsporangia develop on the undersurface of the microsporophyll.

1. Microsporophyll
2. Microsporangia

Figure 6.117 A megasporangiate cone of *Cycas revoluta* showing ovules on leaf–like megasporophylls near the time of pollination.

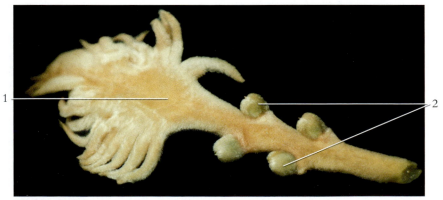

Figure 6.118 The megasporophyll and ovules of *Cycas revoluta*.

1. Megasporophyll   2. Ovules

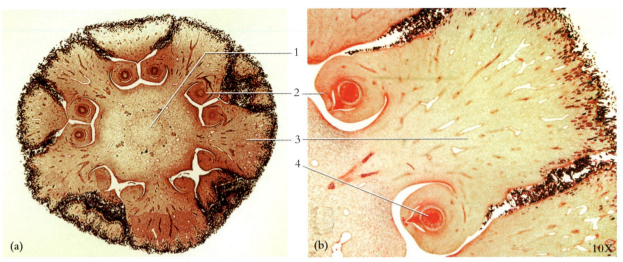

Figure 6.119 Transverse sections of a megasporangiate cone of the cycad *Zamia* sp. (a) A low magnification, and (b) a magnified view.

1. Cone axis
2. Ovule
3. Megasporophyll
4. Megasporocyte

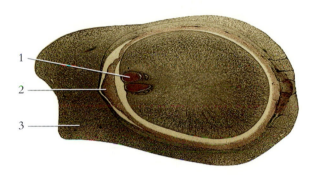

Figure 6.120 An ovule of the cycad *Zamia* sp. The ovule has two archegonia and is ready to be fertilized.

1. Archegonium
2. Megasporangium (nucellus)
3. Integument (will become seed coat)

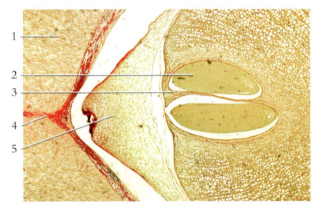

Figure 6.121 A magnified view of the ovule of the cycad *Zamia* sp. showing eggs in archegonia.

1. Integument
2. Egg
3. Archegonium
4. Micropyle area
5. Megasporangium

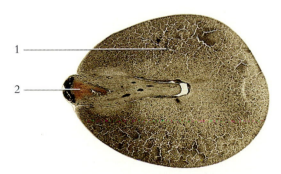

Figure 6.122 An ovule of the cycad *Zamia* sp. The ovule has been fertilized and contains an embryo. The seed coat has been removed from this specimen.

1. Female gametophyte
2. Embryo

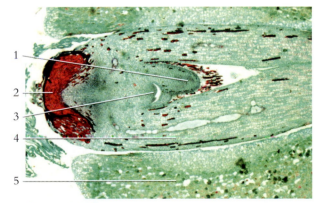

Figure 6.123 A magnified view of the ovule of the cycad *Zamia* sp. showing the embryo.

1. Leaf primordium
2. Root apex
3. Shoot apex
4. Cotyledon
5. Female gametophyte

Ginkgophyta - ginkgo

Figure 6.124 A ginkgo, or maidenhair tree, *Ginkgo biloba*. Consisting of a central trunk with lateral branches, a mature *Ginkgo* grows to 100 meters tall. Native to China, *Ginkgo biloba* has been introduced into countries with temperate climates throughout the world as an interesting and hardy ornamental tree.

Figure 6.126 A fossil *Ginkgo biloba* leaf impression from Paleocene sediment. This specimen was found in Morton County, North Dakota (scale in mm).

Figure 6.128 A branch of a *Ginkgo biloba* tree supporting a mature seed.

  1. Long shoot   3. Mature seed
  2. Short shoot (spur)

Figure 6.125 A leaf from the *Ginkgo biloba* tree. The fan-shaped leaf is characteristic of this species.

Figure 6.127 As the sole member of the phylum Ginkgophyta, *Ginkgo biloba* is able to withstand air pollution. Ginkos are often used as ornamental trees within city parks. *Ginkgo biloba* may have the longest genetic lineage among seed plants.

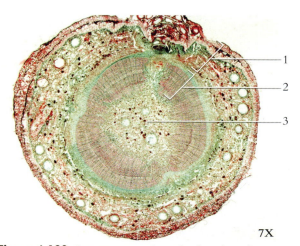

7X

Figure 6.129 A transverse section of a short branch from *Ginkgo biloba*.

  1. Cortex   3. Pith
  2. Vascular tissue

Figure 6.130 The leaves and immature ovules on a short shoot of the ginkgo tree, *Ginkgo biloba*.

1. Leaf 3. Short shoot
2. Immature ovules 4. Long shoot

Figure 6.131 The pollen strobili of the ginkgo tree, *Ginkgo biloba*.

1. Leaf 3. Pollen strobilus
2. Long shoot

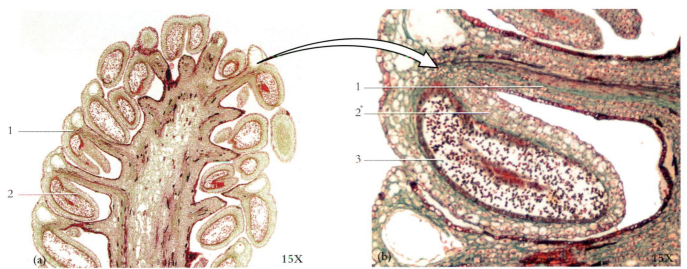

Figure 6.132 A microsporangiate strobilus of *Ginkgo biloba*. (a) A longitudinal section and (b) a magnified view showing a microsporangium.

1. Sporophyll 2. Microsporangium 3. Pollen

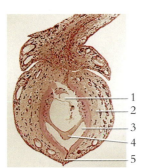

Figure 6.133 A longitudinal section of an ovule of *Ginkgo biloba* prior to fertilization.

1. Megagametophyte
2. Integument
3. Pollen chamber
4. Nucellus
5. Micropyle

Figure 6.134 A transverse and longitudinal sections through a living immature seed of *Ginkgo biloba* showing the green megagametophyte.

1. Fleshy layer of integument
2. Megagametophyte
3. Stoney layer of integument

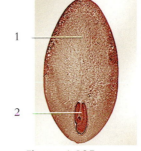

Figure 6.135 A longitudinal section of a seed of *Ginkgo biloba* with the seed coat removed.

1. Megagametophyte
2. Developing embryo

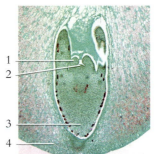

Figure 6.136 A magnified view of the ovule of *Ginkgo biloba* showing the embryo.

1. Leaf primordium
2. Shoot apex
3. Root apex
4. Megagametophyte

Pinophyta (=Coniferae) - conifers

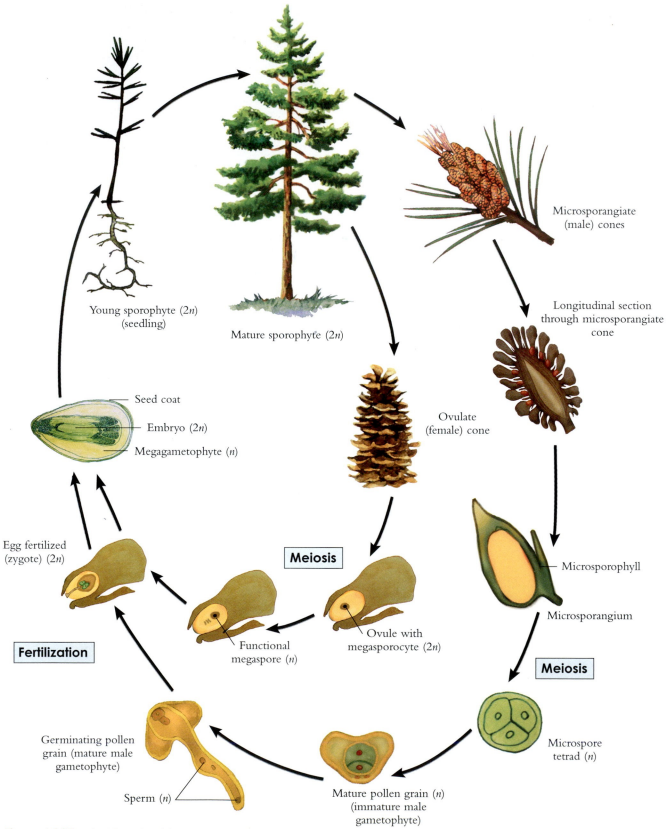

Microsporangiate (male) cones

Longitudinal section through microsporangiate cone

Young sporophyte (2n) (seedling)

Mature sporophyte (2n)

Seed coat

Embryo (2n)

Megagametophyte (n)

Ovulate (female) cone

Microsporophyll

Microsporangium

Meiosis

Egg fertilized (zygote) (2n)

Meiosis

Functional megaspore (n)

Ovule with megasporocyte (2n)

Fertilization

Microspore tetrad (n)

Germinating pollen grain (mature male gametophyte)

Sperm (n)

Mature pollen grain (n) (immature male gametophyte)

Figure 6.137   The life cycle of the pine, *Pinus* sp.

Courtesy of Champion Paper Company, Inc.

Figure 6.138 A diagram of the tissues in the stem (trunk) of a conifer. The periderm and dead secondary phloem (outer bark) protects the tree against water lost and the infestation of insects and fungi. The cells of the phloem (inner bark) compress and become nonfunctional after a relatively short period. The vascular cambium annually produces new phloem and xylem and accounts for the growth rings in the wood. The secondary xylem is a water-transporting layer of the stem and provides structural support to the tree.

1. Outer bark
2. Phloem
3. Vascular cambium
4. Secondary xylem

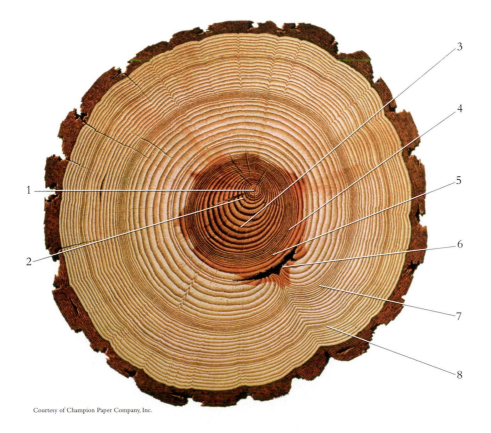

Courtesy of Champion Paper Company, Inc.

Figure 6.139 The stem (trunk) of a pine tree that was harvested in the year 2000 when the tree was 62 years old. The growth rings of a tree indicate environmental conditions that occurred during the tree's life.

1. **1939**—A pine seedling.
2. **1944**—Healthy, undisturbed growth indicated by broad and evenly spaced rings.
3. **1949**—Growth disparity probably due to the falling of a dead tree onto the young healthy six-year-old tree. The wider "reaction rings" on the lower side help support the tree.
4. **1959**—The tree is growing straight again, but the narrow rings indicate competition for sunlight and moisture from neighboring trees.
5. **1962**—The surrounding trees are harvested, thus permitting rapid growth once again.
6. **1965**—A burn scar from a fire that quickly scorched the forest.
7. **1977**—Narrow growth rings resulting from a prolonged drought.
8. **1992**—Narrow growth rings resulting from a sawfly insect infestation, whose larvae eat the needles and buds of many kinds of conifers.

(a) (b) (c)

Figure 6.140 The leaves of most species of conifers are needle-shaped such as those of the blue spruce, *Picea pungens* (a). *Araucaria heterophyla,* Norfolk Island pine, however, (b) has awl-shaped leaves, and *Podocarpus* sp. (c) has strap-shaped leaves.

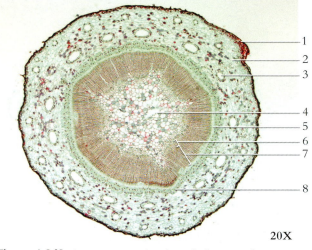

20X

Figure 6.141 A transverse section through the stem of a young conifer showing the arrangement of the tissue layers.

1. Epidermis	5. Cambium
2. Cortex	6. Primary xylem
3. Resin duct	7. Spring wood of secondary xylem
4. Pith	8. Primary phloem

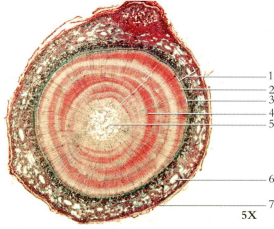

5X

Figure 6.142 A transverse section through the stem of *Pinus* sp., showing secondary stem growth.

1. Bark (cortex, and periderm)	4. Secondary xylem
2. Secondary phloem	5. Pith
3. Vascular cambium	6. Resin duct
	7. Epidermis

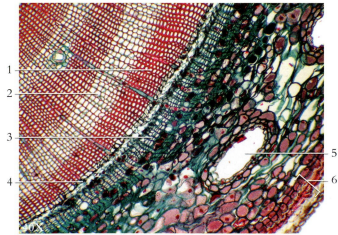

40X

Figure 6.143 An enlarged view of the stem of *Pinus* sp. showing tissues following secondary growth.

1. Late secondary xylem (wood)	4. Vascular cambium
2. Early secondary xylem (wood)	5. Resin duct
3. Secondary phloem	6. Periderm

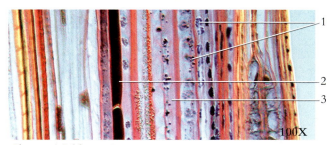

100X

Figure 6.144 A radial longitudinal section through the phloem of *Pinus* sp.

1. Sieve areas on a sieve cell	3. Sieve cell
2. Storage parenchyma	

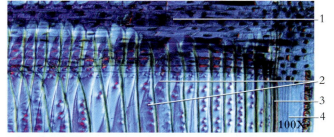

100X

Figure 6.145 A radial longitudinal section through a stem of *Pinus* sp., cut through the xylem tissue.

1. Ray parenchyma	3. Vascular cambium
2. Tracheids	4. Sieve cells

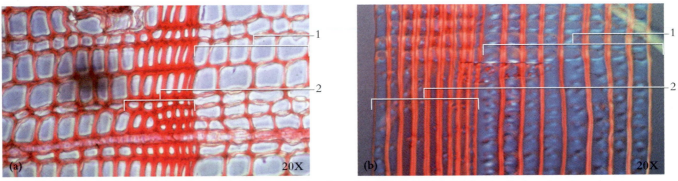

Figure 6.146 The growth rings in *Pinus* sp. (a) Transverse section through a stem; and (b) radial longitudinal section through a stem.
1. Early wood   2. Late wood

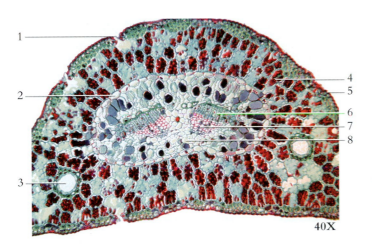

Figure 6.147 The transverse section of a leaf (needle) of *Pinus* sp.

1. Stoma
2. Endodermis
3. Resin duct
4. Photosynthetic mesophyll
5. Epidermis
6. Phloem
7. Xylem
8. Transfusion tissue

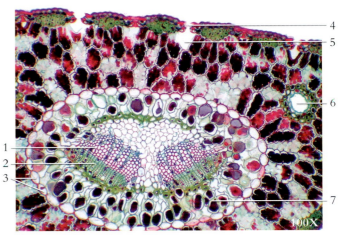

Figure 6.148 The transverse section through the leaf (needle) of *Pinus* sp.

1. Xylem
2. Phloem
3. Endodermis
4. Sunken stoma
5. Sub-stomatal chamber
6. Resin duct
7. Transfusion tissue (surrounding vascular tissue)

Aibes sp.   *Taxodium* sp.   *Araucaria* sp.   *Taxus* sp.   *Pinus* sp.

Figure 6.149 The megasporangiate cones from various species of conifers.

Figure 6.150 A first-year ovulate cone in *Pinus* sp.
1. Pollen cones   2. First-year ovulate cone

Figure 6.51 A transverse section through a first-year ovulate cone in *Pseudotsuga* sp. (scale in mm).
1. Immature ovules   2. Cone scale bracts

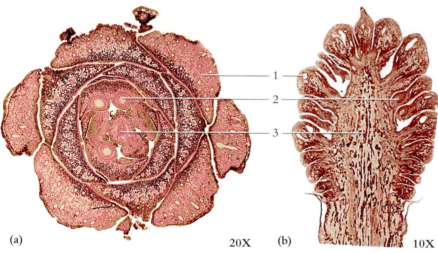

(a)   20X   (b)   10X

Figure 6.152 An ovulate cones of a *Pinus* sp. (a) Transverse section, and (b) longitudinal section.
1. Ovuliferous scale   2. Ovule   3. Cone axis

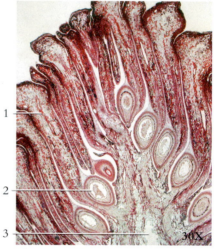

30X

Figure 6.153 A magnified view of a *Pinus* sp. ovulate cone (longitudinal view).
1. Ovuliferous scale   3. Cone axis
2. Ovule

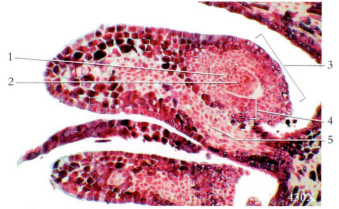

430X

Figure 6.154 A magnified view of a *Pinus* sp. ovule (immature).
1. Megaspore mother cell   3. Ovule
2. Nucellus   4. Integument
        5. Cone scale

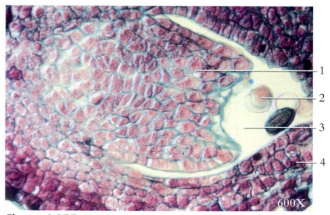

600X

Figure 6.155 A magnified view of an ovule of *Pinus* sp. with pollen grains in the pollen chamber.
1. Nucellus   3. Pollen chamber
2. Pollen grain   4. Integument

Figure 6.156 Microsporangiate cones of (a) *Pinus* sp. prior to the release of pollen and (b) *Picea pungens* after pollen has been released. The pollen cones are at the end of a branch.

   1. Needle-like leaves   2. Microsporophylls   3. Pollen cone

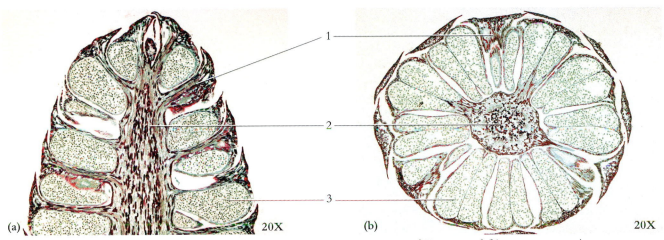

Figure 6.157 (a) Longitudinal section through the tip of a microsporangiate cone of *Pinus* sp. and (b) a transverse section.

   1. Sporophyll   2. Cone axis   3. Microsporangium

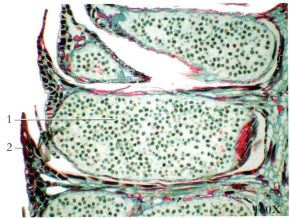

Figure 6.158 A close-up of a microsporangiate cone scale and microsporangium of *Pinus* sp.

   1. Microsporangium with   2. Microsporophyll
       pollen grains

Figure 6.159 A micrograph of stained pollen grains of *Pinus* sp., showing wings.

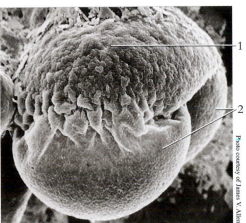

Figure 6.160 a scanning electron micrograph of a *Pinus* sp. pollen grain with inflated bladder-like wings.

   1. Pollen body   2. Wings

Photo courtesy of James V. Allen

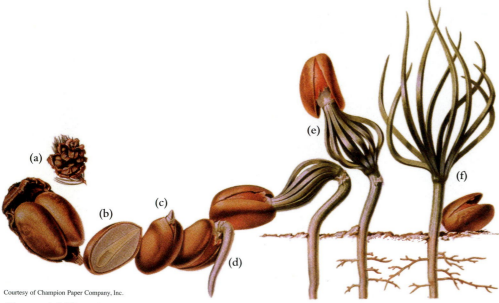

Courtesy of Champion Paper Company, Inc.

Figure 6.161 A diagram of pinyon pine seed germination producing a young sporophyte. (a) The seeds are protected inside the cone, two seeds formed on each scale. (b) A sectioned seed shows an embryo embedded in the female gametophyte tissue. (c) The growing embryo splits the shell of the seed, enabling the root to grow toward the soil. (d) As soon as the tiny root tip penetrates and anchors into the soil, water and nutrients are absorbed. (e) The cotyledons emerge from the seed coat and create a supply of chlorophyll. Now the sporophyte can manufacture its own food from water and nutrients in the soil and carbon dioxide in the air. (f) Growth occurs at the terminal buds at the base of the leaves.

Figure 6.162 A young sporophyte (seedling) of a pine, *Pinus* sp. (scale in mm).
1. Seedling leaves (needles)
2. Young stem
3. Young roots

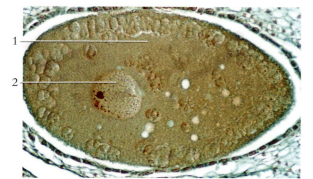

Figure 6.163 A close-up of an ovulate cone scale in *Pinus* sp.
1. Mature seeds (wings)
2. Ovulate cone scale
3. Seed (containing embryo within seed coat)

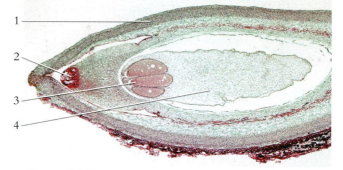

Figure 6.164 A young ovule of *Pinus* sp. showing the megaagmetophyte.
1. Ovule
2. Micropyle
3. Archegonium
4. Megagametophyte

Figure 6.165 A young ovule of *Pinus* sp. showing the egg in archegonium.
1. Egg
2. Nucleus

Figure 6.166 A magnified view of the ovule of *Pinus* sp. showing the embryo.
1. Integument
2. Micropyle
3. Leaf primordium
4. Root primordium

Magnoliophyta (=Anthophyta) – Flowering Plants

Monocots

One cotyledon

Flower parts in threes or multiples of three

Leaf veins parallel

Vascular bundles scattered

Some examples of monocots

Wheat

Corn

Cattail

Iris

Dicots

Two cotyledons

Flower parts in fours or fives or multiples of four or five

Leaf veins form a net pattern

Vascular bundles arranged in a ring

Some examples of dicots

Water lily

Columbine

Rose

Sunflower

Figure 6.167 A comparison and examples of monocots and dicots.

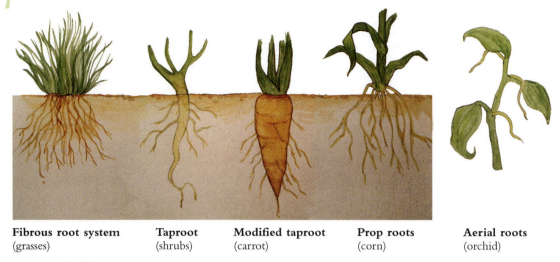

Fibrous root system
(grasses)

Taproot
(shrubs)

Modified taproot
(carrot)

Prop roots
(corn)

Aerial roots
(orchid)

Figure 6.168 The root systems of angiosperms.

(a)

(b)

Figure 6.169 The root system of an orchid (monocot) (a) showing aerial roots and a banyan (dicot) (b) showing prop roots. Monocot roots are fibrous, with many roots of more or less equal size. Dicots usually have a taproot system, consisting of a long central root with smaller, secondary roots branching from it.

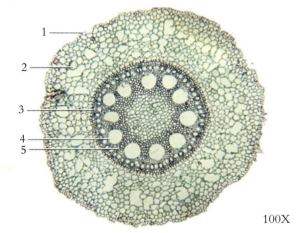

100X

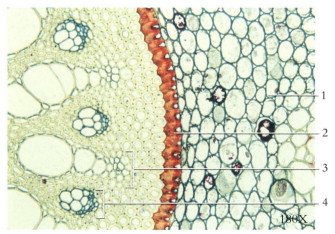

180X

Figure 6.170 A transverse section of the root of the monocot *Smilax* sp.

1. Epidermis   4. Xylem
2. Cortex   5. Phloem
3. Endodermis

Figure 6.171 A close-up of a root of the monocot *Smilax* sp.

1. Cortex   2. Endodermis
3. Xylem   4. Phloem

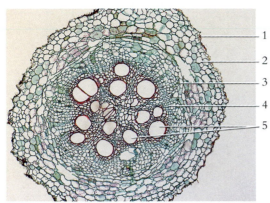

Figure 6.172 A transverse section of a sweet potato root, *Ipomaea* sp.

1. Remnants of epidermis
2. Cortex
3. Endodermis
4. Phloem
5. Xylem

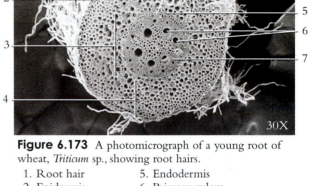

Figure 6.173 A photomicrograph of a young root of wheat, *Triticum* sp., showing root hairs.

1. Root hair
2. Epidermis
3. Stele
4. Cortex
5. Endodermis
6. Primary xylem
7. Primary phloem

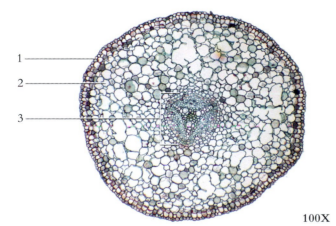

Figure 6.174 A longitudinal section of a willow species showing lateral root formation.

1. Lateral root
2. Epidermis
3. Cortex
4. Vascular tissue

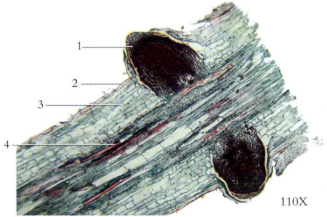

Figure 6.175 A transverse section branch root formation of *Phaseolus* sp.

1. Epidermis
2. Cortex
3. Branch root
4. Vascular tissue (stele)

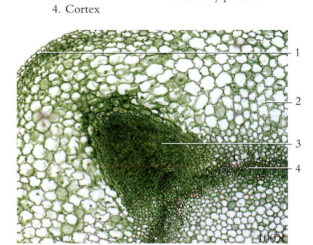

Figure 6.176 A transverse section of a young root of *Salix* sp.

1. Epidermis
2. Cortex
3. Stele

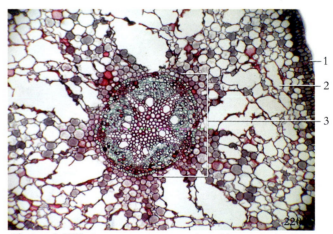

Figure 6.177 A transverse section of an older root of *Salix* sp., showing early secondary growth.

1. Epidermis
2. Cortex
3. Vascular tissue

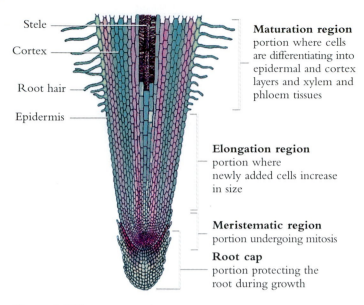

Stele

Cortex

Root hair

Epidermis

Maturation region portion where cells are differentiating into epidermal and cortex layers and xylem and phloem tissues

Elongation region portion where newly added cells increase in size

Meristematic region portion undergoing mitosis

Root cap portion protecting the root during growth

Figure 6.178 A diagram of a root tip.

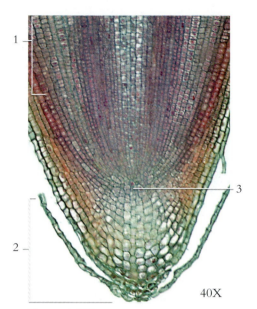

40X

Figure 6.179 A photomicrograph of the root tip of a pear, *Pyrus* sp., seen in longitudinal section.
1. Elongation region   3. Apical meristem
2. Root cap

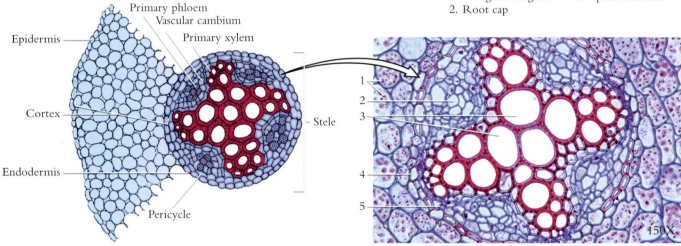

Primary phloem
Vascular cambium
Primary xylem

Epidermis

Cortex

Endodermis

Stele

Pericycle

150X

Figure 6.180 (a) A diagram of a transverse section of a dicot root. (b) A photomicrograph showing a transverse section of the stele.
1. Starch grains within parenchyma cells   3. Primary xylem   5. Pericycle
2. Primary phloem   4. Endodermis

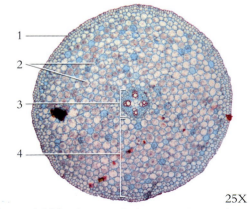

25X

Figure 6.181 The root of a buttercup, *Ranunculus* sp.
1. Epidermis   3. Stele
2. Parenchyma cells of cortex   4. Cortex

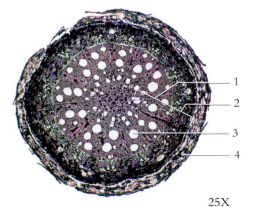

25X

Figure 6.182 A transverse section of the root of basswood, *Tilia* sp., showing secondary growth.
1. Secondary xylem   3. Vessel element
2. Secondary phloem   4. Periderm

Figure 6.183 Some examples of the variety and specialization of angiosperm stems. The stem of an angiosperm is often the ascending portion of the plant specialized to produce and support leaves and flowers, transport and store water and nutrients, and provide growth through cell division. Stems of plants are utilized extensively by humans in products including paper, building materials, furniture, and fuel. In addition, the stems of potatoes, onions, cabbage, and other plants are important food crops.

Figure 6.184 An example of specialized underground stems. (a) A potato, and (b) an onion.
  1. Node (eye) bearing a minute scale leaf and stem bud
  2. Bulb scales (modified leaves)
  3. Short stem

Figure 6.185 The woody stem of a dicot seen in early spring just as the buds are beginning to swell. Branches and twigs are small extensions of the stems of angiosperms and often support leaves and flowers.
  1. Terminal (apical) bud
  2. Internode
  3. Terminal bud scale scars
  4. Lenticel
  5. Lateral (axillary) bud
  6. Node
  7. Leaf (vascular bundle) scar

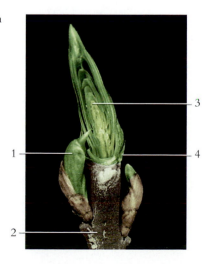

Figure 6.186 The terminal bud of a woody stem that has been longitudinally sectioned to show developing leaves.
  1. Lateral (axillary) bud
  2. Stem
  3. Leaf primordia
  4. Bud scale

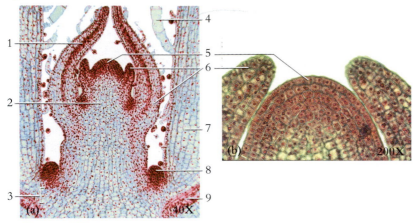

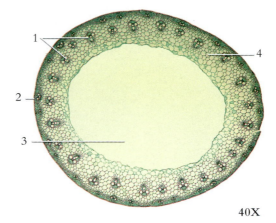

Figure 6.187 A longitudinal section of the stem tip of the common houseplant *Coleus* sp.

1. Procambium
2. Ground meristem
3. Leaf gap
4. Trichome
5. Apical meristem
6. Developing leaf primordia
7. Leaf primordium
8. Axillary bud
9. Developing vascular tissue

Figure 6.188 A transverse section through the stem of a monocot, *Triticum* sp., wheat.

1. Vascular bundles
2. Epidermis
3. Ground tissue cavity
4. Parenchyma cells

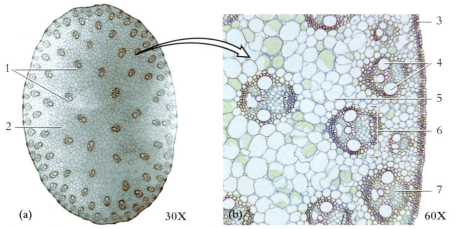

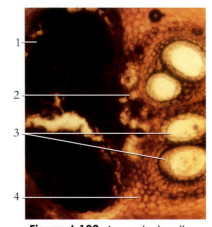

Figure 6.189 A transverse section from the stem of a monocot, *Zea mays* (corn).

1. Vascular bundles with primary xylem and phloem
2. The pattern of vascular bundles in a monocot, is known as an atactostele
3. Epidermis
4. Vessel elements of primary xylem
5. Parenchyma cells
6. Vascular bundle
7. Primary phloem

Figure 6.190 A vascular bundle of a fossil palm plant.

1. Bundle cap (fibers)
2. Phloem
3. Vessel elements
4. Ground tissue (parenchyma)

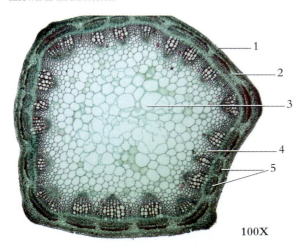

Figure 6.191 A transverse section through a stem of clover, *Trifolium* sp. showing an eustele.

1. Epidermis
2. Cortex
3. Pith
4. Interfascicular region
5. Vascular bundles with caps of phloem fibers

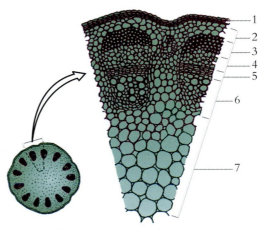

Figure 6.192 A diagram of vascular bundles from the stem of a dicot showing the eustele.

1. Early periderm
2. Cortex
3. Phloem fibers
4. Phloem
5. Vascular cambium
6. Xylem
7. Pith

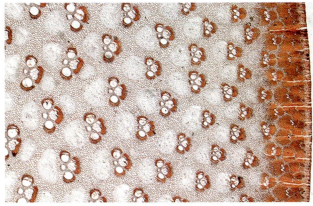

"Woody" monocot

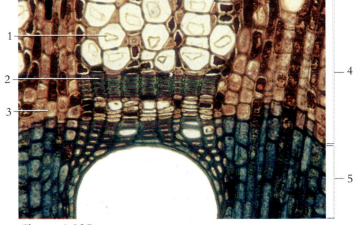

Woody dicot

Figure 6.193 A comparison of the transverse sections of stems of a "woody" monocot and a woody dicot. The stem of the "woody" monocot is rigid because of the fibrous nature of the numerous vascular bundles. The stem of the woody dicot is rigid because of the compact xylem cells impregnated with lignin forming the dense, hardened wood, seen as annual rings.

   1. Bark   2. Annual rings

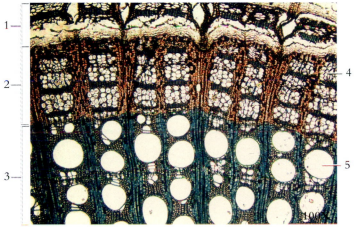

Figure 6.194 A transverse section of a *Vitis* sp. stem showing secondary tissues.

   1. Bark   4. Sieve tube elements
   2. Secondary phloem   5. Vessel
   3. Secondary xylem

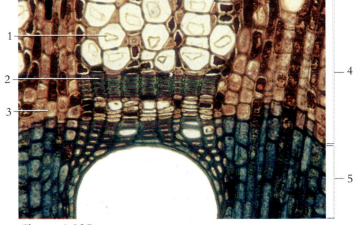

Figure 6.195 A transverse section of a *Vitis* sp. stem.

   1. Sieve tube element   4. Secondary phloem
   2. Phloem fibers   5. Secondary xylem
   3. Parenchyma cells

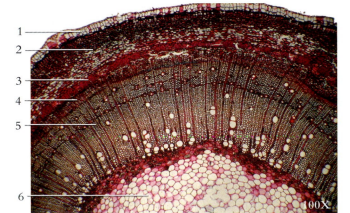

Figure 6.796 A transverse section through one-year-old *Fraxinus* sp. stem showing secondary growth.

   1. Periderm   4. Secondary phloem
   2. Cortex   5. Secondary xylem
   3. Phloem fibers   6. Pith

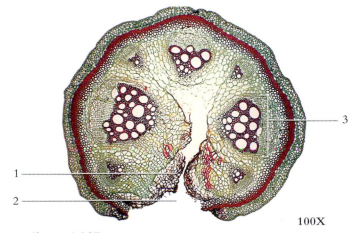

Figure 6.197 A stem of *Aristolochia* sp. with healing wound.

   1. Callus tissue   3. Vascular bundle
   2. Wound

(a)   (b)   (c)   (d)   (e)   (f)

Figure 6.198 Some samples of bark patterns of representative conifers and angiosperms

(a) **Redwood**—The tough, fibrous bark of a redwood tree may be 30 cm thick. It is highly resistant to fire and insect infestation.

(b) **Ponderosa pine**—The mosaic-like pattern of the bark of mature ponderosa pine is resistant to fire.

(c) **White birch**—The surface texture of bark on the white birch is like white paper. The bark of the white birch was used by Indians in Eastern United States for making canoes.

(d) **Sycamore**—The mottled color of the sycamore bark is due to a tendency for large, thin, brittle plates to peel off, revealing lighter areas beneath. These areas grow darker with exposure, until they, too, peel off.

(e) **Mangrove**—The leathery bark of a mangrove tree is adaptive to brackish water in tropical or semi-tropical regions.

(f) **Shagbark hickory**—The strips of bark in a mature shagbark hickory tree give this tree its common name.

Figure 6.199 An angiosperm, *Ruscus aculeatus*, is characterized by stems (a) that resemble leaves in form and function. Note the true leaf (b) arising from the leaf-like stem.

1. Stem   2. Leaf   3. Flower bud

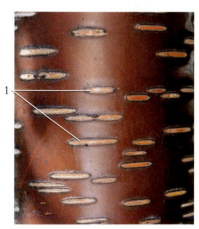

Figure 6.200 The bark of a birch tree, *Betula occidentalis*, showing lenticels. Lenticels are spongy areas in the cork surfaces that permit gas exchange between the internal tissues and the atmosphere.

1. Lenticels

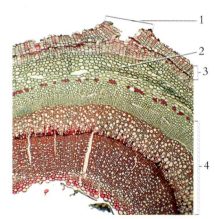

Figure 6.201 A transverse section of a dicot stem showing a lenticel and stem tissues.

1. Lenticel   3. Periderm
2. Cortex   4. Vascular tissue

Figure 6.202 A gall on an oak, *Quercus* sp., stem. The feeding of a gall wasp larva causes abnormal growth and the formation of a gall. The wasp larva feeds upon the gall tissue, pupates within this enclosure, and then chews an exit to emerge.

1. Gall   2. Stem

Venation	**Margin**	**Complexity**	**Arrangement on Stem**
Pinnate	Entire	Palmately compound	Opposite
Parallel	Pinnately lobed	Simple	Alternate
Palmate	Serrate	Pinnately compound	Whorled

Figure 6.203 Several representative angiosperm leaf types. Leaves comprise the foliage of plants, which provide habitat and a food source for many animals including humans. Leaves also provide protective ground cover and are the portion of the plant most responsible for oxygen replenishment into the atmosphere.

(a)   (b)   (c)

Figure 6.204 The shape of the leaf (a) is of adaptive value to withstand wind. As the speed of the wind increases (b) and (c), the leaf rolls into a tight cone shape, avoiding damage.

Figure 6.205 An angiosperm leaf showing characteristic surface features. Leaves are organs modified to carry out photosynthesis. Photosynthesis is the manufacture of food (sugar) from carbon dioxide and water, with sunlight providing energy.

1. Lamina (blade)   3. Midrib   5. Petiole
2. Serrate margin   4. Veins

Figure 6.206 The undersurface of an angiosperm leaf showing the vascular tissue lacing through the lamina, or blade, of the leaf.

1. Midrib
2. Secondary veins

Figure 6.207 The organic decomposition of a leaf is a gradual process beginning with the softer tissues of the lamina, leaving only the vascular tissues of the midrib and the veins, as seen in this photograph. With time, these will also decompose.

Figure 6.208 Some examples of specialized leaves for floatation. (a) Leaves from a giant water lily. (b) Water hyacinths, *Eichhornia* sp., have modified leaves that buoy the plants on the water surface. Water hyacinths are common in New World tropical fresh-water habitats, where they may become so thick that they choke out bottom-dwelling plants and clog waterways.

Figure 6.209 As seen on the leaflets in the upper right of this photograph, the leaves of the sensitive plant, *Mimosa pudica*, droop upon being touched. The drooping results from differential changes in turgor of the leaf cells in the pulvinus, a thickened area at the base of the leaflet.

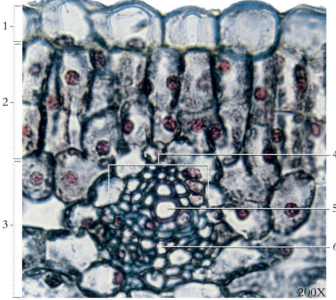

Figure 6.210 A transverse section of tomato leaf, *Lycopersicon* sp.

1. Upper epidermis   4. Leaf vein (vascular bundle)
2. Palisade mesophyll   5. Xylem
3. Spongy mesophyll   6. Phloem

Figure 6.211 A transverse section through the leaf of the common hedge privet *Ligustrum* sp. The typical tissue arrangement of a leaf includes an upper epidermis, a lower epidermis, and the centrally located mesophyll. Containing chloroplasts, the cells of the mesophyll are often divided into palisade mesophyll and spongy mesophyll. Veins within the mesophyll conduct material through the leaf.

1. Upper epidermis   5. Xylem
2. Palisade mesophyll   6. Phloem
3. Gland   7. Spongy mesophyll
4. Bundle sheath   8. Lower epidermis

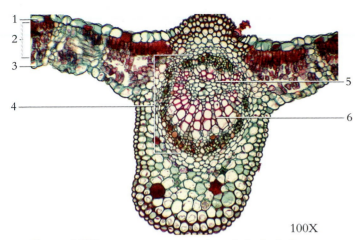

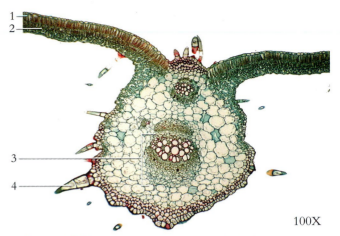

Figure 6.212 A transverse section through the leaf of basswood, *Tilia* sp.
1. Upper epidermis   4. Leaf vein (midrib)
2. Mesophyll   5. Phloem
3. Lower epidermis   6. Xylem

Figure 6.213 A transverse section through the leaf of cucumber, *Cucurbita* sp.
1. Palisade mesophyll   3. Leaf vein (midrib)
2. Spongy mesophyll   4. Trichome

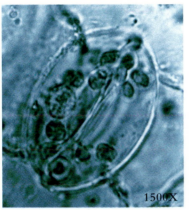

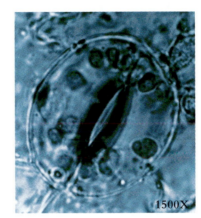

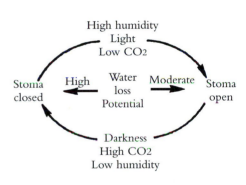

High humidity
Light
Low CO2

Stoma   High   Water   Moderate   Stoma
closed   loss   open
               Potential

Darkness
High CO2
Low humidity

Figure 6.214 The guard cells in many plants regulate the opening of the stomata according to the environmental factors, as indicated in this diagram. (a) Face view of a closed stoma of a geranium, and (b) an open stoma.

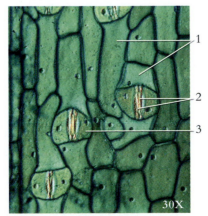

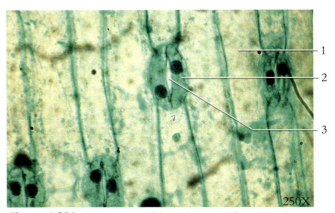

Figure 6.215 A surface view of the leaf epidermis of *Tradescantia* sp.
1. Epidermal cells
2. Guard cells surrounding stomata
3. Subsidiary cells

Figure 6.216 A face view of the epidermis of onion, *Allium* sp. Note the twin guard cells with the stoma opened.
1. Lower epidermis   3. Stoma
2. Guard cell

Flowers of Angiosperms

Structure of a flower

Pistil
— Stigma
— Style
— Ovary
— Ovule

Anther
Filament
Stamen

Petal

Sepal

Receptacle

Pedicel

Position of ovaries

Hypogynous
(superior ovary)

Perigynous
(centrally positioned ovary)

Epigynous
(inferior ovary)

Figure 6.217 Diagrams of angiosperm flowers showing the structure and relative position of the ovaries. The angiosperm flower is typically composed of sepals, petals, stamens, and one or more pistils. The sepals are the outermost circle of protective leaf-like structures. They are usually green and are collectively called the calyx. The stamens and the pistils are the reproductive parts of the flower. A stamen consists of the filament (stalk) and the anther, where pollen is produced. The pistil consists of a sticky stigma at the tip that receives pollen and a style that leads to the ovary.

Crown of Thorns

Foxglove

Orchid

Bird of Paradise

Columbine

Hibiscus

Figure 6.218 Some flowers of representative angiosperms.

Figure 6.219 The floral structure of a tulip, *Tulipa* sp.

1. Petal
2. Anther
3. Stigma
4. Filament
5. Style

Figure 6.220 The structure of a dissected cherry, *Prunus* sp., showing a perigynous flower.

1. Petal
2. Filaments
3. Sepal
4. Anther
5. Stigma
6. Style
7. Floral tube

Figure 6.221 The structure of a dissected pear, *Pyrus* sp., showing an epigynous flower.

1. Petal
2. Anther
3. Filament
4. Style
5. Sepal
6. Ovary

Figure 6.222 A dissected quince, *Chaenomeles japonica*, showing an epigynous flower.

1. Petal
2. Anther
3. Stigma
4. Filament
5. Style
6. Ovules

(a)   (b)   (c)

Figure 6.223 (a) The floral structure of *Gladiolus* sp. (b) The anthers and stigma and (c) the ovary.

1. Anther
2. Filament
3. Ovules
4. Receptacle
5. Stigma
6. Style
7. Ovary
8. Anther
9. Pollen
10. Stigma
11. Style
12. Filament
13. Ovules (immature seeds)
14. Receptacle
15. Style
16. Ovary

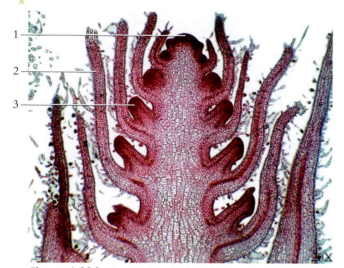

Figure 6.224 The floral bud of Coleus, *Coleus* sp.

1. Apical meristem   3. Floral bud
2. Bract

Figure 6.225 The ovary of tomato, *Lycopersicon* sp., with developing ovules

1. Ovary wall   3. Placenta
2. Ovules

Figure 6.226 A nightshade, *Solanum* sp., floral bud showing ovary with developing ovules.

1. Ovary wall   3. Placenta
2. Ovules

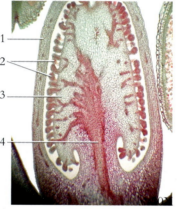

Figure 6.227 The floral bud of tobacco, *Nicotiana* sp., showing the ovary and ovules.

1. Ovary wall   3. Placenta
2. Ovules   4. Vascular tissue

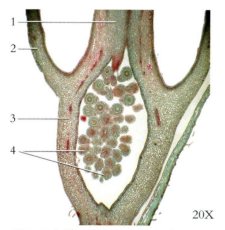

Figure 6.228 The floral bud of a currant, *Ribes* sp., showing an inferior ovary with developing ovules.

1. Style   3. Ovary
2. Petal   4. Ovules

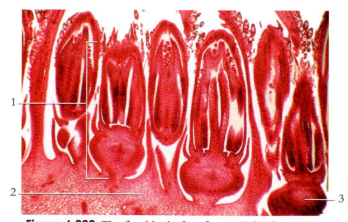

Figure 6.229 The floral bud of sunflower, *Helianthus* sp., with several immature flowers.

1. Individual flower   3. Ovary of individual flower
2. Receptacle

Figure 6.230 A scanning electron micrograph of the stigma of an angiosperm pistil. The stigma is the location where pollen grains adhere and germinate to produce a pollen tube.

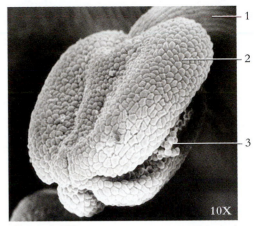

Figure 6.231 A scanning electron micrograph of the anther of candy tuft, *Lobularia* sp. The anther has ruptured, resulting in the release of pollen grains.

1. Filament   3. Pollen grains
2. Anther

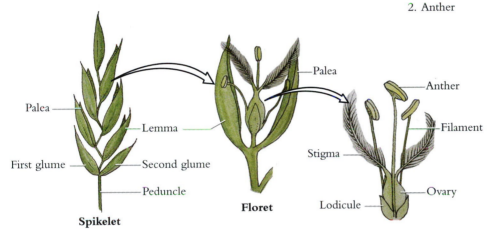

Figure 6.232 The floral structure of grasses.

Palea

Lemma

First glume   Second glume

Peduncle

Spikelet

Palea

Floret

Anther

Filament

Stigma

Lodicule   Ovary

Figure 6.233 The floral parts of a grass, *Elymus flavescens,* showing spikelets with six florets.

Figure 6.234 The grass, wheat, *Triticum* sp. This species is economically important.

Figure 6.235 Bamboo an important grass of commerce. It is also important in many natural ecosystems.

Figure 6.236 Corn, *Zea mays,* a New World grass that is important as food for humans and livestock.

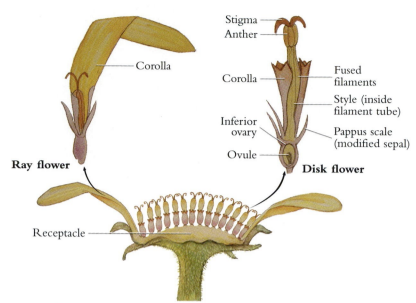

Stigma
Anther
Corolla
Corolla
Fused filaments
Style (inside filament tube)
Inferior ovary
Pappus scale (modified sepal)
Ovule
Ray flower
Disk flower
Receptacle

Figure 6.237 The flowers of the family Asteraceae are usually produced in tight heads resembling single large flowers. One of these inflorescences can contain hundreds of individual flowers. Examples of this family include dandelions, sunflowers, asters, and marigolds.

Figure 6.238 A dissected inflorescence of a member of the Asteraceae, *Balsamorhiza sagittata.*
1. Ray flower   3. Receptacle
2. Disk flower

Figure 6.239 A strawberry, *Fragaria* sp., showing (a) the flower, (b) immature aggregate fruits, and (c) a ripening fruit.

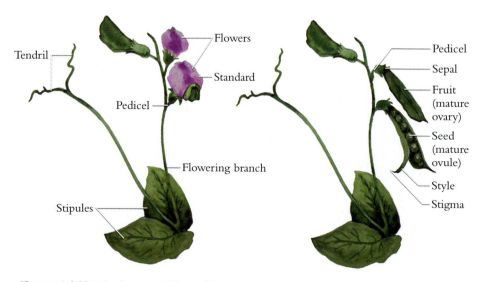

Tendril
Flowers
Standard
Pedicel
Flowering branch
Stipules

Pedicel
Sepal
Fruit (mature ovary)
Seed (mature ovule)
Style
Stigma

Figure 6.240 The flower and fruit of the pea, *Pisum* sp.

Figure 6.241 The seeds on the receptacle of the giant sunflower.

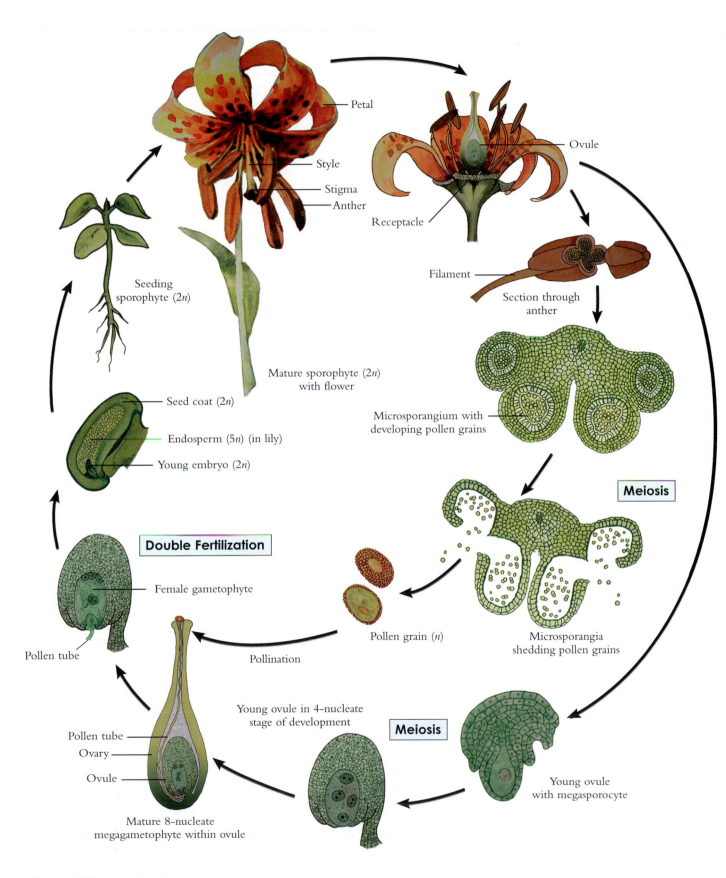

Figure 6.242 The life cycle of an angiosperm.

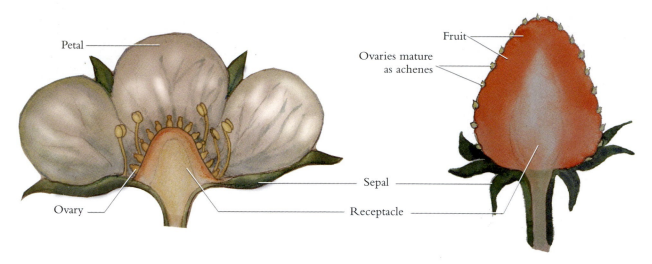

Figure 6.243 The flower and fruit of the strawberry, *Fragaria* sp. The strawberry is an aggregate fruit.

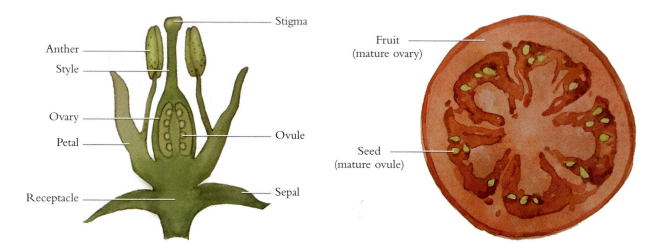

Figure 6.244 An illustration of a flower and fruit of a tomato, *Lycopersicon esculentum*. A tomato fruit is a berry.

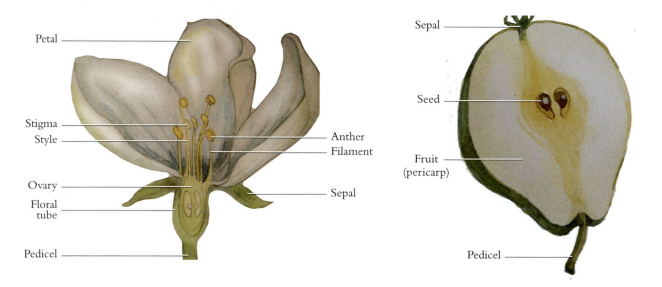

Figure 6.245 The flower and fruit of the pear *Pyrus* sp. The pear fruit develops from the floral tube (fused perianth) as well as the ovary.

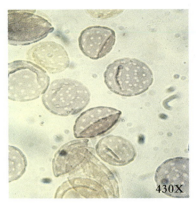

Figure 6.246 The pollen grains of the dicot pigweed, *Amaranthus* sp.

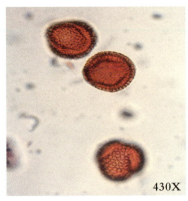

Figure 6.247 The pollen grains of a lilac, *Syringa* sp.

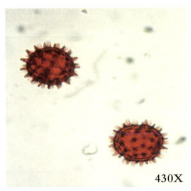

Figure 6.248 The pollen grains of the dicot arrowroot, *Balsamorhiza* sp.

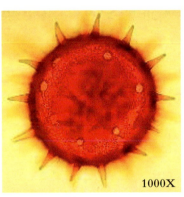

Figure 6.249 The pollen grain of hibiscus, *Hibiscus* sp.

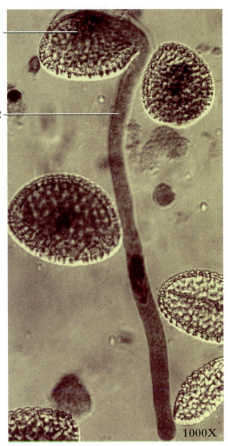

1
2

Figure 6.250 The pollen grains of a lily. The pollen grain at the top of the photo has germinated to produce a pollen tube.
1. Pollen grain   2. Pollen tube

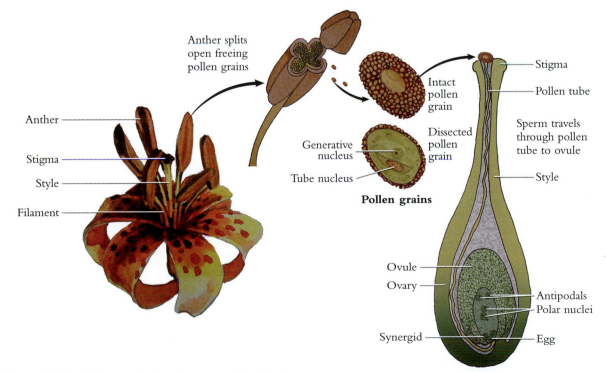

Figure 6.251 A diagram showing the process of pollination.

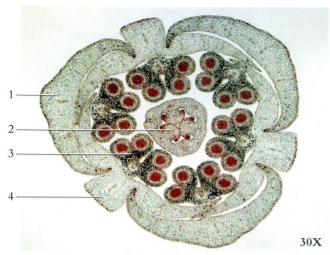

Figure 6.252 A transverse section of a flower bud from a lily, *Lilium* sp.
1. Sepal   3. Anther
2. Ovary   4. Petal

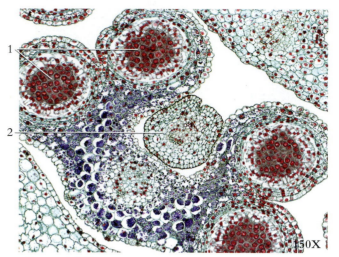

Figure 6.253 A transverse section of an anther from a lily, *Lilium* sp.
1. Sporogenous tissue   2. Filament

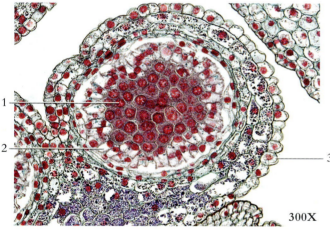

Figure 6.254 A transverse section of an anther from a lily, *Lilium* sp.
1. Young microsporocytes   2. Tapetum   3. Anther wall

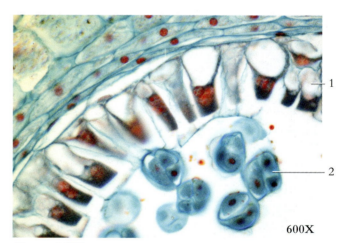

Figure 6.255 A transverse section of an anther from a lily, *Lilium* sp., magnified view.
1. Tapetum   2. Tetrad of microspores

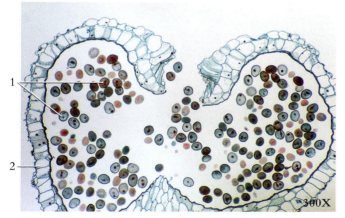

Figure 6.256 A transverse section of an anther from a lily, *Lilium* sp., showing mature pollen.
1. Pollen grains with two cells   2. Anther wall

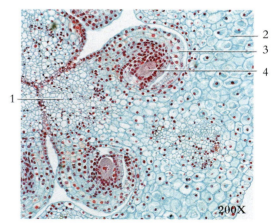

Figure 6.257 A transverse section of a lily, *Lilium* sp., ovary showing ovules.
1. Placenta   3. Ovule
2. Ovary wall   4. Megasporocyte ($2n$)

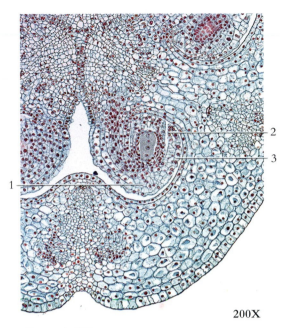

200X

Figure 6.258 A transverse section of a lily, *Lilium* sp., ovary showing megaspore.
1. Ovule
2. Linear tetrad of megaspore
3. Integument

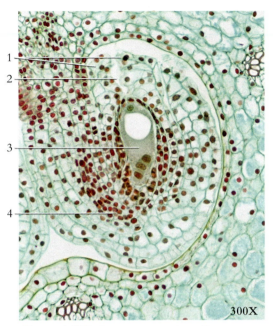

300X

Figure 6.259 A transverse section of a lily, *Lilium* sp., ovary showing ovule with developing embryo sac.

1. Integuments   3. Embryo sac
2. Micropyle   4. Ovule

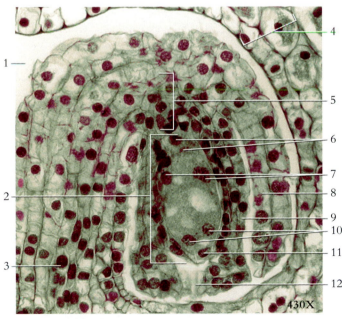

430X

Figure 6.260 A transverse section of an eight-nucleate embryo sac of an ovule from a lily, *Lilium* sp.
1. Locule
2. Megagametophyte
3. Funiculus
4. Wall of ovary
5. Chalaza
6. Antipodal cells ($3n$)
7. Polar nuclei ($3n$)
8. Outer integument ($2n$)
9. Inner integument ($2n$)
10. Synergid cells (n)
11. Egg (n)
12. Micropyle (pollen tube entrance)

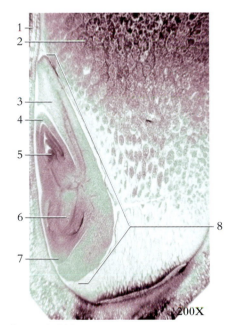

200X

Figure 6.261 A photomicrograph of a mature grain, or kernel, of wheat, *Triticum aestivum*.
1. Pericarp   5. Shoot apex
2. Starchy endosperm   6. Radicle
3. Scutellum   7. Coleorhiza
4. Coleoptile   8. Embryo

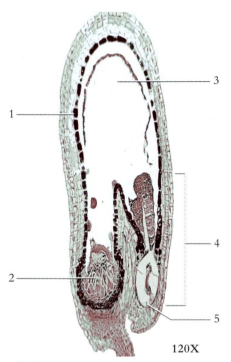

120X

Figure 6.262 A photomicrograph of a developing dicot embryo from a shepherd's purse, *Capsella bursa-pastoris*.

1. Endothelium
2. Cellular endosperm
3. Endosperm
4. Developing embryo
5. Basal cell

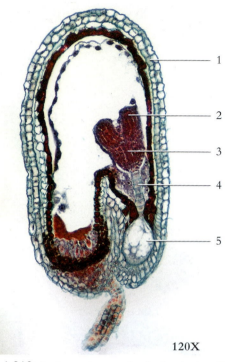

120X

Figure 6.263 A photomicrograph of a developing dicot embryo from a shepherd's purse, *Capsella bursa-pastoris*, showing young embryo.

1. Seed coat
2. Cotyledon
3. Hypocotyl
4. Suspensor
5. Basal cell

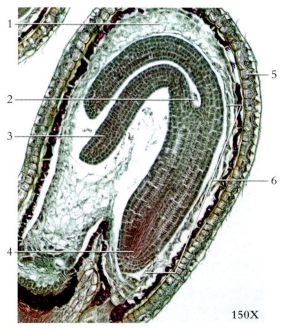

150X

Figure 6.264 A photomicrograph of a developing dicot embryo from a shepherd's purse, *Capsella bursa-pastoris*, showing a nearly mature embryo.

1. Endosperm
2. Epicotyl
3. Cotyledon
4. Radicle
5. Seed coat
6. Hypocotyl

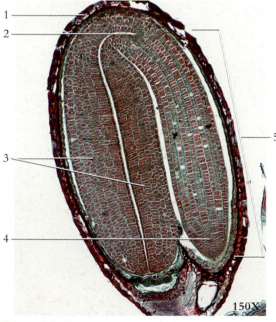

150X

Figure 6.265 A photomicrograph of a developing dicot embryo from a shepherd's purse, *Capsella bursa-pastoris*, showing a mature embryo..

1. Seed coat
2. Epicotyl
3. Cotyledons
4. Radicle
5. Hypocotyl

(a)

(b)

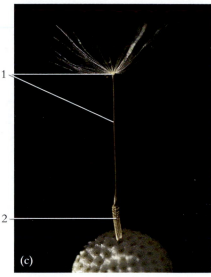
1

2

(c)

Figure 6.266 The flower (a) and the fruits (b and c) of the dandelion, *Taraxacum* sp. The dandelion has a composite flower. The wind-borne fruit (containing one seed) of a dandelion, and many other members of the family Asteraceae, develop a plume-like pappus, which enables the light fruit to float in the air.

1. Pappus   2. Ovary wall, with one seed inside

Figure 6.267 A dissected legume, garden bean, *Phaseolus* sp.
1. Pedicel   3. Fruit
2. Seeds   4. Style

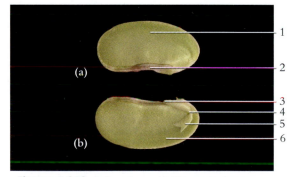

Figure 6.268 A lima bean. (a) The entire bean seed and (b) a longitudinally sectioned seed
1. Integument (seed coat)   4. Hypocotyl
2. Hilum   5. Epicotyl (plume)
3. Radicle   6. Cotyledon

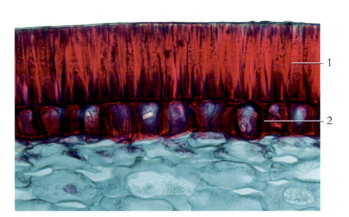

Figure 6.269 A photomicrograph of the seed coat of the garden bean, *Phaseolus* sp., showing the sclerified epidermis
1. Macrosclereids   2. Subepidermal sclereids

Figure 6.270 A cob of corn from *Zea mays*. Corn was domesticated approximately 7,000 years ago from a Mexican grass, family Poaceae.

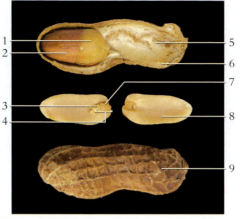

Figure 6.271 The fruit and seed of a peanut plant.

1. Cotyledon
2. Integument (seed coat)
3. Plumule
4. Embryo axis
5. Interior of fruit
6. Mesocarp
7. Radicle
8. Cotyledon
9. Fruit wall (pericarp)

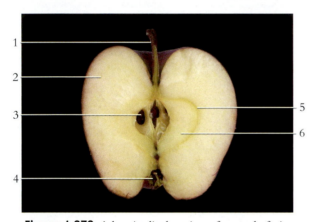

Figure 6.272 A longitudinal section of an apple fruit.

1. Pedicel
2. Mature floral tube
3. Seed (mature ovule)
4. Remnants of floral parts
5. Ovary wall
6. Mature ovary (2 & 6 comprise the fruit)

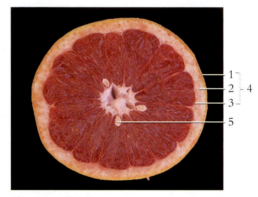

Figure 6.273 A transverse section through a grapefruit fruit.

1. Exocarp
2. Mesocarp
3. Endocarp
4. Pericarp
5. Seed

Figure 6.274 A longitudinal section of a pineapple fruit.

1. Shoot apex   2. Central axis   3. Floral parts

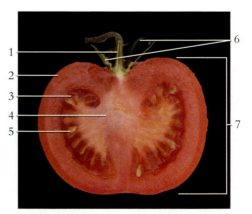

Figure 6.275 A longitudinal section of a tomato fruit (berry).

1. Pedicel
2. Pericarp
3. Locule
4. Placenta
5. Seed
6. Sepals
7. Mature ovary (fruit)

(a)   (b)   (c)   (d)   (e)   (f)

(g)   (h)   (i)   (j)   (k)   (l)

Figure 6.276 Some examples of seed dispersal.

(a) Maple—The winged fruits of a maple fall with a spinning motion that may carry it hundreds of yards from the parent tree.
(b) White pine—The second-year cones of a white pine open to expose the winged seeds to the wind.
(c) Willow—The airborne seeds of a willow may be dispersed over long distances.
(d) Witch hazel—Mature seeds of the witch hazel tree are dispersed up to 10 feet by forceful discharge.
(e) Mangrove—The fruits of this tropical tree begin to germinate while still on the branch, forming pointed roots. When the seeds drop from the tree, they may float to a muddy area where the roots take hold.
(f) Coconut—The buoyant, fibrous husk of a coconut permits dispersal from one island or land mass to another by ocean currents.

(g) Pecan—The fruit husk of a pecan provides buoyancy and protection as it is dispersed by water.
(h) Black walnut—The encapsulated seed of the black walnut is dispersed through burial by a squirrel or floating in a stream.
(i) Apple—The seeds of an apple tree may be dispersed by animals that ingest the fruit and pass the undigested seeds hours later in their feces.
(j) Cherry—Moderate-sized birds, such as robins, may carry a ripe cherry to an eating site where the juicy pulp is eaten and the hard seed is discarded.
(k) Beech—Seeds from a beech tree are dispersed by mammals as the spiny husks adhere to their hair. In addition, many mammals ingest these seeds and disperse them in their feces.
(l) Oak—An oak seed may be dispersed through burial of the acorn fruit by a squirrel or jay.

Figure 6.277 (a) The mature milkweed, *Asclepias* sp.; (b) milkweed pods; and (c) seeds ready for airborne dispersal.

Forcible discharge dispersal

Touch-me-not

Water dispersal

Coconut

Animal dispersal

Burdock

Cocklebur

Wind dispersal

Dandelion

Poppy

Blackberries

Maple

Figure 6.278 Several fruits and seeds to illustrate seed dispersal.

Animals are multicellular, heterotrophic eukaryotes that ingest food materials and store carbohydrate reserves as glycogen or fat. The cells of animals lack cell walls but do contain intercellular connections including desmosomes, gap junctions, and tight junctions. Animal cells are also highly specialized into the specific kinds of tissues described in chapter 1. Most animals are motile through the contraction of muscle fibers containing actin and myosin proteins. The complex body systems of animals include elaborate sensory and neuromotor specializations that accommodate dynamic behavioral mechanisms.

Reproduction in animals is primarily sexual, with the diploid stage generally dominating the life cycle. Primary sex organs, or *gonads*, produce the haploid gametes called *sperm* and *egg*. Propagation begins as a small flagellated sperm fertilizes a larger, nonmotile egg, forming a diploid zygote that has genetic traits of both parents. The zygote then undergoes a succession of mitotic divisions called *cleavage*. In animals, cleavage is followed by the formation of a multicellular stage called a *blastula*. With further development, the *germ layers* form, which eventually give rise to each of the body organs. The developmental cycle of many animals includes *larval forms*, which are still developing, free-living, and sexually immature. Larvae usually have food and habitat requirements different from those of the adults. Larvae eventually undergo metamorphoses, which transform them into sexually mature adults.

Animals inhabit nearly all aquatic and terrestrial habitats of the biosphere. The greatest number of animals are marine, where the first animals probably evolved. Depending on the classification scheme, animals may be grouped into as many as 35 phyla. The most commonly known phylum is *Chordata* (table 7.1), which includes the subphylum *Vertebrata*, or the backboned animals. Chordates, however, comprise only about 5% of all the animal species. All other animals are frequently referred to as *invertebrates*, and they account for approximately 95% of the animal species.

Table 7.1 Some Representatives of the Kingdom Animalia

Phylum and Representative Kinds	Characteristics
Porifera — sponges	Multicellular, aquatic animals, with stiff skeletons and bodies perforated by pores
Cnidaria — corals, hydra, and jellyfish	Aquatic animals, radially symmetrical, mouth surrounded by tentacles bearing cnidocytes (stinging cells); body composed of epidermis and gastrodermis, separated by mesoglea
Platyhelminthes — flatworms	Elongated, flattened, and bilaterally symmetrical; distinct head containing ganglia; nerve cords; protonephridia or flame cells
Mollusca — clams, snails, and squids	Bilaterally symmetrical with a true coelom, containing a mantle; many have muscular foot and protective shell
Annelida — segmented worms	Body segmented (except leeches); a series of hearts; hydrostatic skeleton and circular and longitudinal muscles
Nematoda — round worms	Mostly microscopic, unsegmented worm-like; body enclosed in cuticle; whip-like body movement
Arthropoda — crustaceans, insects, and spiders	Body segmented; paired and jointed appendages; chitinous exoskeleton; hemocoel for blood flow
Echinodermata — sea stars and sea urchins	Larvae have bilateral symmetry; adults have pentaradial symmetry; coelom; most contain a complete digestive tract; regeneration of body parts
Chordata — lancelets, tunicates, and vertebrates	Fibrous notochord, pharyngeal gill slits, dorsal hollow nerve cord, and postanal tail present at some stage in their development

Phylum Porifera

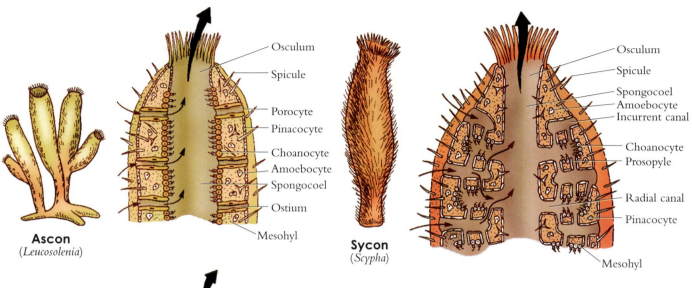

Ascon
(*Leucosolenia*)

Osculum
Spicule
Porocyte
Pinacocyte
Choanocyte
Amoebocyte
Spongocoel
Ostium
Mesohyl

Sycon
(*Scypha*)

Osculum
Spicule
Spongocoel
Amoebocyte
Incurrent canal
Choanocyte
Prosopyle
Radial canal
Pinacocyte
Mesohyl

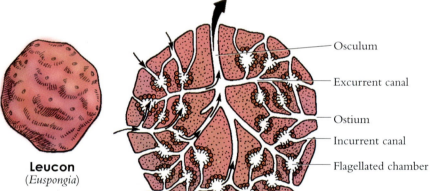

Leucon
(*Euspongia*)

Osculum
Excurrent canal
Ostium
Incurrent canal
Flagellated chamber

Figure 7.1 Examples of sponge body types. A diagrammatic representative of each of the three types depicts with arrows the flow of water through the body of the sponge.

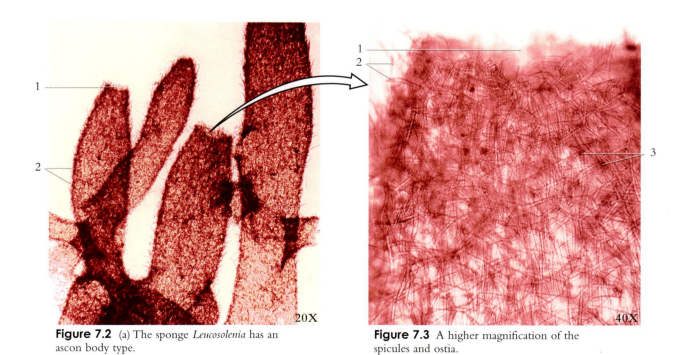

Figure 7.2 (a) The sponge *Leucosolenia* has an ascon body type.
1. Osculum   2. Spicules

Figure 7.3 A higher magnification of the spicules and ostia.
1. Osculum   2. Spicules   3. Ostia

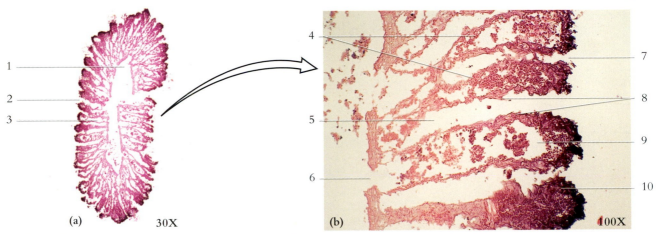

(a)   30X

(b)   100X

Figure 7.4 Transverse sections of the sponge, *Grantia* (*Scypha*). (a) Low magnification and (b) high magnification.

1. Spongocoel
2. Incurrent canal
3. Radial canal
4. Choanocytes (collar cells)
5. Incurrent canal
6. Apopyle
7. Ostium
8. Pinacocytes
9. Radial canal
10. Mesohyl

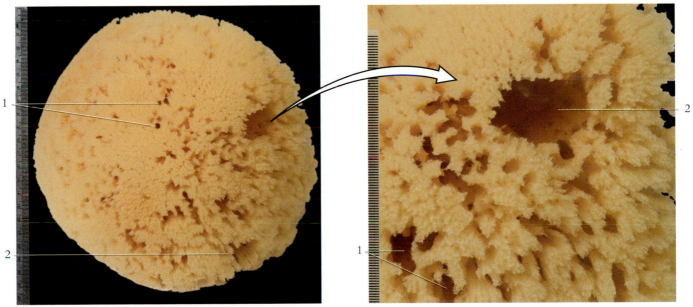

Figure 7.5 A bath sponge, class Demospongiae, has a leuconoid body structure (scale in mm).

1. Ostia   2. Osculum

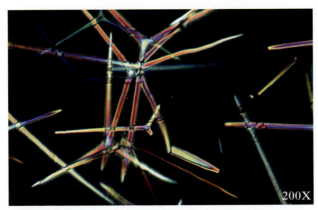

Figure 7.6 The branched silica spicules of a freshwater sponge.

Figure 7.7 A yellow ball sponge, *Cinachyra allocladia* a leucon-type sponge (scale in mm).

Phylum Cnidaria

Table 7.2 Representatives of the Phylum Cnidaria

Classes and Representative Kinds	Characteristics
Hydrozoa — hydra, *Obelia*, and Portuguese man-of-war	Mainly marine; both polyp and medusa stage (polyps form only in hydra); polyp colonies in most
Scyphozoa — jellyfish	Marine coastal waters; polyp stage restricted to small larval forms
Cubozoa — box jellies	Marine coastal waters; polyp and medusa stage; square-shaped when viewed from above
Anthozoa — sea anemones, corals, and sea fans	Marine coastal waters; solitary or colonial polyps; no medusa stage; partitioned gastrovascular cavity

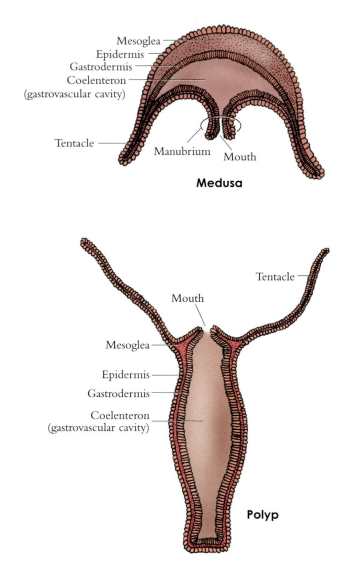

Figure 7.8 The general body type of cnidarians.

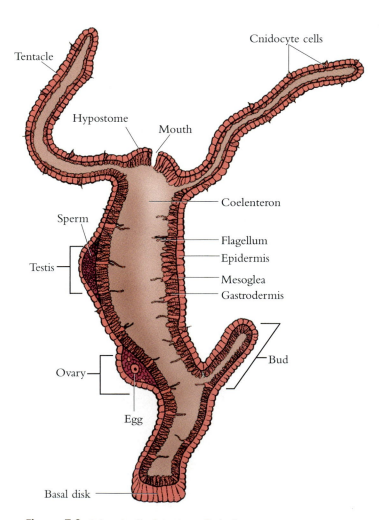

Figure 7.9 A longitudinal section of a hydra.

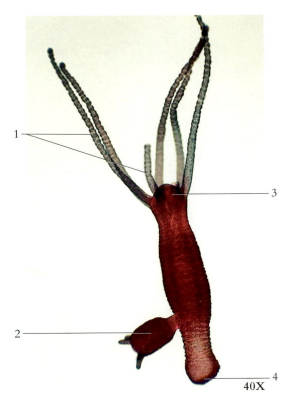

Figure 7.10 A budding hydra.

1. Tentacles
2. Bud
3. Hypostome
4. Basal disk (foot)

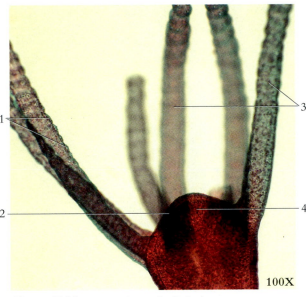

Figure 7.11 An anterior end of a hydra.

1. Cnidocytes
2. Hypostome
3. Tentacles
4. Mouth

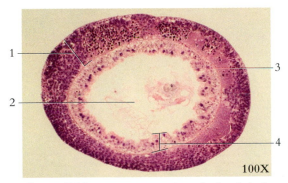

Figure 7.12 A transverse section of a female hydra.

1. Epidermis
2. Coelenteron
3. Mesoglea
4. Gastrodermis

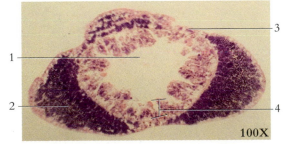

Figure 7.13 A transverse section of a male hydra.

1. Coelenteron
2. Testes
3. Epidermis
4. Gastrodermis

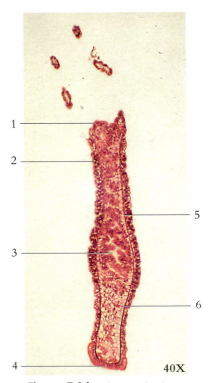

Figure 7.14 A longitudinal section of a hydra.

1. Hypostome
2. Epidermis
3. Coelenteron
4. Basal disk
5. Mesoglea
6. Gastrodermis

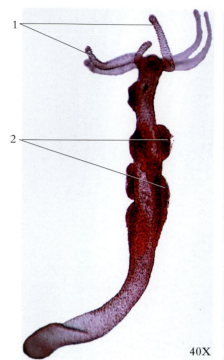

Figure 7.15 A male hydra.
1. Tentacles
2. Testes

Figure 7.16 A female hydra.
1. Tentacles
2. Ovary
3. Basal disk (foot)

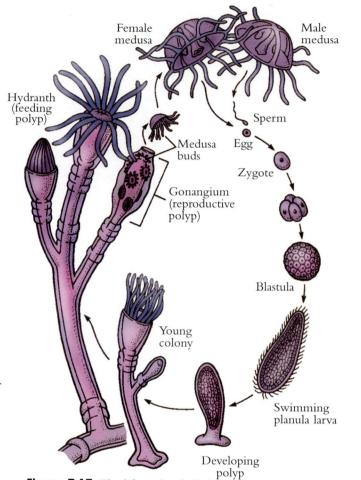

Figure 7.17 The life cycle of *Obelia*.

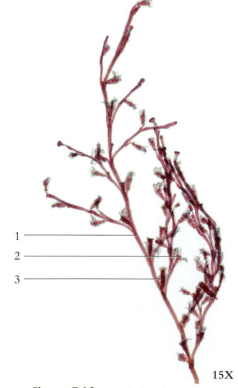

Figure 7.18 An *Obelia* colony.
1. Coenosarc (soft tissue connecting polyps)
2. Hydranth (feeding polyp)
3. Gonangium (reproductive polyp)

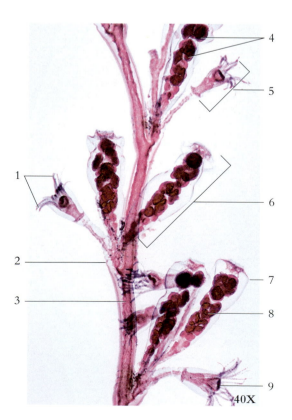

Figure 7.19 A detail of an *Obelia* colony.
1. Tentacles
2. Perisarc (horny covering that encloses the polyp)
3. Coenosarc
4. Medusa buds
5. Hydranth (feeding polyp)
6. Gonangium (reproductive polyp)
7. Gonotheca
8. Blastostyle
9. Hypostome

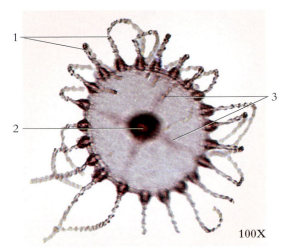

Figure 7.20 *Obelia* medusa.
1. Tentacles   3. Radial canals
2. Manubrium

Figure 7.21 *Obelia* medusa in feeding position.
1. Tentacles   3. Manubrium
2. Gonad   4. Mouth

Figure 7.22 The Portuguese man-of-war, *Physalia* sp. It is a colony of medusae and polyps acting as a single organism. The tentacles are composed of three types of polyps: the gastrozooids (feeding polyps), the dactylozooids (stinging polyps), and the gonozooids (reproductive polyps) (scale in mm).
1. Pneumatophore (float)   2. Tentacles

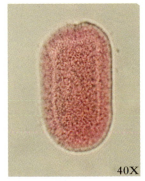

Figure 7.23 The *Aurelia* planula larva. It develops from a fertilized egg that may be retained on the oral arm of the medusa.

Figure 7.24 An *Aurelia* scyphistoma. The polyp is a developmental stage in the life cycle of the jellyfish.

Figure 7.25 An *Aurelia* strobila. Under favorable conditions, the scyphistoma develops into the strobila.
1. Developing ephyrae

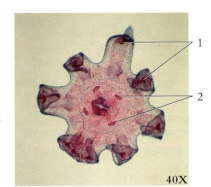

Figure 7.26 An *Aurelia* ephyra larva. It gradually develops into adult jellyfish.
1. Rhopalia (sense organs)
2. Gonads

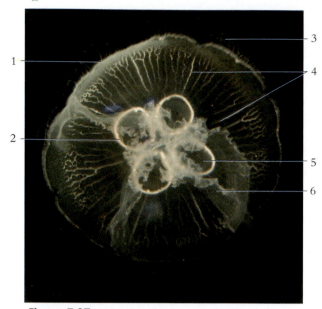

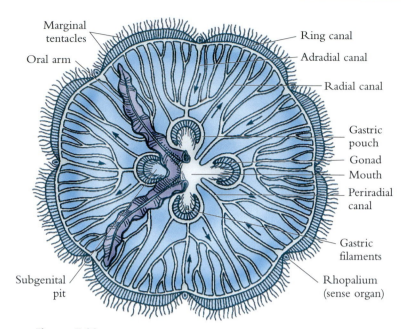

Figure 7.27 An oral view of *Aurelia* medusa.

1. Ring canal   3. Marginal tentacles   5. Subgenital pit
2. Gonad   4. Radial canals   6. Oral arm

Figure 7.28 An oral view of *Aurelia* medusa. In this diagram, the right oral arms have been removed. The arrows depict circulation through the canal system.

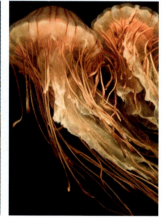

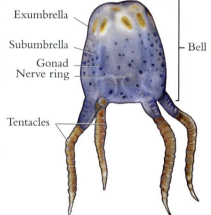

Figure 7.29 The sea nettle, *Chrysaora fuscescens*. They often mass in large swarms off the Pacific coast, where they feed on zooplankton.

Figure 7.30 The red-striped jellyfish, *Chrysaora melanaster*, are common near the surface of the Bering sea.

Figure 7.31 A box jellyfish, *Carybdea sivickisi*, named for their cube-shaped bell. All cubozoans have four tentacles.

Figure 7.32 An illustration of box jellyfish, *Carybdea sivickisi*, showing basic structures.

Figure 7.33 The sunburst anemone, *Anthopleura sola*. Its green coloration comes from symbiotic algae within it.

Figure 7.34 Two examples of anemone, disk anemones and a yellow anemone.

Figure 7.35 The sea pen, *Ptilosarcus gurneyi*.

Figure 7.36 The disk anemone, *Actinodiscus*. It forms large colonies.

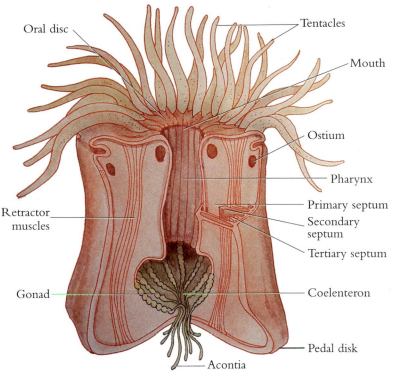

Oral disc

Tentacles

Mouth

Ostium

Pharynx

Primary septum

Secondary septum

Tertiary septum

Retractor muscles

Gonad

Coelenteron

Pedal disk

Acontia

Figure 7.37 A diagram of a partially dissected sea anemone, *Metridium*.

Figure 7.38 Brain coral, *Goniastrea*.

Figure 7.39 The skeletal structure of brain coral, *Goniastrea*.

Figure 7.40 Mushroom coral, *Rhodactis*.

Figure 7.41 The skeletal structure of mushroom coral, *Rhodactis*.

Figure 7.42 Staghorn coral, *Acropora*.

Figure 7.43 The skeletal structure of staghorn coral, *Acropora*.

Figure 7.44 A detailed view of the polyps of candy cane coral, *Caulastrea furcata*.

Figure 7.45 A detailed view of the polyps of glove xenia, *Xenia umbellata*.

Phylum Platyhelminthes

Table 7.3 Some Representatives of the Phylum Platyhelminthes

Classes and Representative Kinds	Characteristics
Turbellaria — planarians	Mostly free-living, carnivorous, aquatic forms; body covered by ciliated epidermis
Trematoda — flukes (schistosomes)	Parasitic with wide range of invertebrate and vertebrate hosts; suckers for attachment to host
Cestoda — tapeworms	Parasitic in many vertebrate hosts; complex life cycle with intermediate hosts; suckers or hooks on scolex for attachment to host; eggs are produced and shed within proglottids

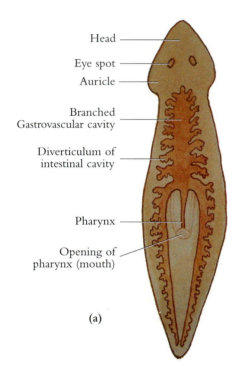

(a)

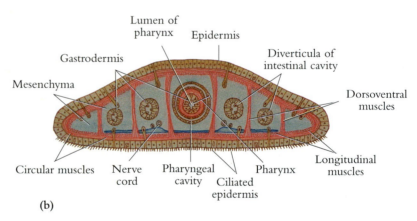

(b)

Figure 7.46 The internal anatomy of *Planaria*. (a) Longitudinal section, and (b) transverse section through the pharyngeal region.

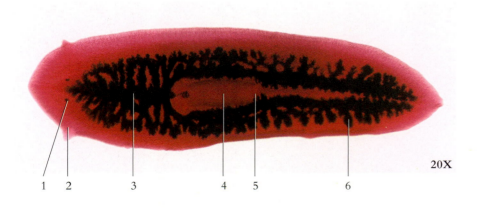

20X

Figure 7.47 *Planaria*.
1. Eye spot
2. Auricle
3. Gastrovascular cavity
4. Pharynx
5. Opening of pharynx (mouth)
6. Diverticulum of intestinal cavity

1   2   3   4   5   6

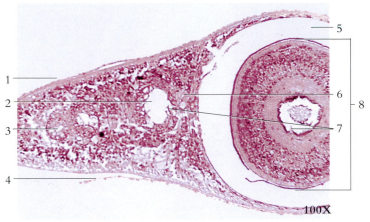

Figure 7.48 A transverse section through the pharyngeal region of *Planaria*.

1. Epidermis
2. Intestinal cavity
3. Testis
4. Cilia
5. Pharyngeal cavity
6. Dorsoventral muscles
7. Gastrodermis
8. Pharynx

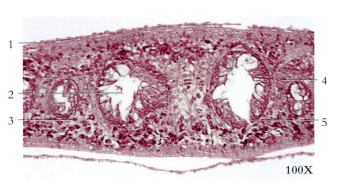

Figure 7.49 A transverse section through the posterior region of *Planaria*.

1. Epidermis
2. Intestinal cavity
3. Mesenchyme
4. Dorsoventral muscles
5. Gastrodermis

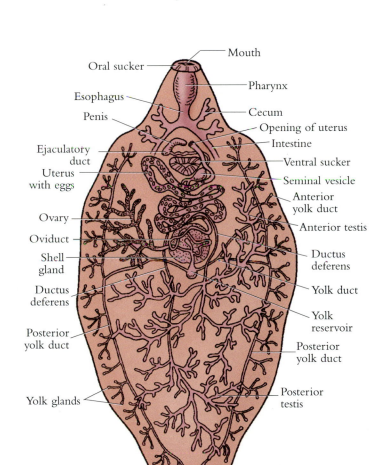

Figure 7.50 A diagram of the sheep liver fluke, *Fasciola hepatica*.

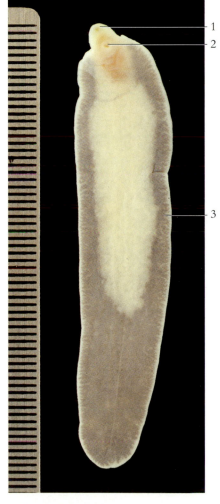

Figure 7.51 The cow liver fluke, *Fasciola magna*, is one of the largest flukes, measuring up to 7.75 cm long (scale in mm).

1. Oral sucker
2. Ventral sucker
3. Yolk gland

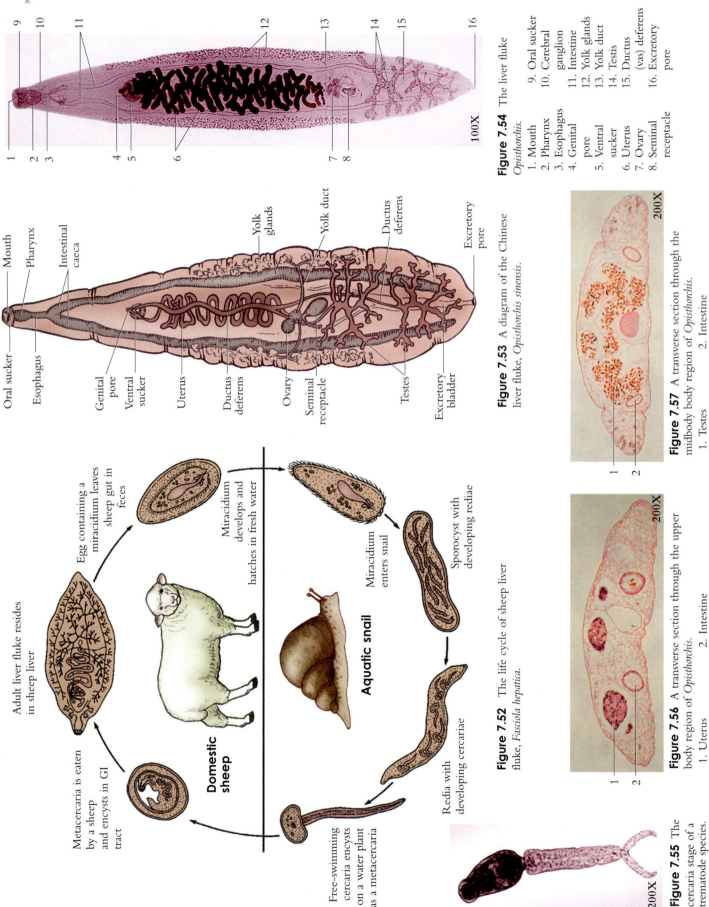

100X

Figure 7.54 The liver fluke *Opisthorchis*.

1. Mouth
2. Pharynx
3. Esophagus
4. Genital pore
5. Ventral sucker
6. Uterus
7. Ovary
8. Seminal receptacle
9. Oral sucker
10. Cerebral ganglion
11. Intestine
12. Yolk glands
13. Yolk duct
14. Testis
15. Ductus (vas) deferens
16. Excretory pore

Oral sucker — Mouth
Pharynx
Esophagus
Intestinal caeca
Genital pore
Ventral sucker
Uterus
Ductus deferens
Ovary
Seminal receptacle
Testes
Excretory bladder
Yolk glands
Yolk duct
Ductus deferens
Excretory pore

Figure 7.53 A diagram of the Chinese liver fluke, *Opisthorchis sinensis*.

200X

Figure 7.57 A transverse section through the midbody region of *Opisthorchis*.
1. Testes   2. Intestine

Egg containing a miracidium leaves sheep gut in feces

Miracidium develops and hatches in fresh water

Miracidium enters snail

Sporocyst with developing rediae

Aquatic snail

Adult liver fluke resides in sheep liver

Domestic sheep

Redia with developing cercariae

Metacercaria is eaten by a sheep and encysts in GI tract

Free-swimming cercaria encysts on a water plant as a metacercaria

Figure 7.52 The life cycle of sheep liver fluke, *Fasciola hepatica*.

200X

Figure 7.56 A transverse section through the upper body region of *Opisthorchis*.
1. Uterus   2. Intestine

200X

Figure 7.55 The cercaria stage of a trematode species.

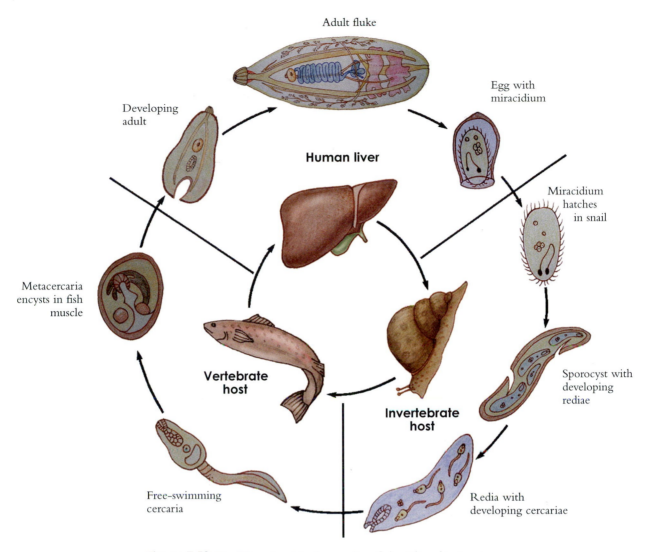

Figure 7.58 The life cycle of the human liver fluke, *Clonorchis sinensis*.

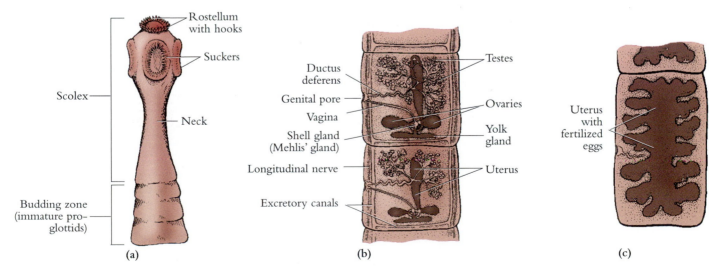

Figure 7.59 Diagrams of a parasitic tapeworm, *Taenia pisiformis*. (a) The anterior end, (b) mature proglottids, and (c) a ripe (gravid) proglottid.

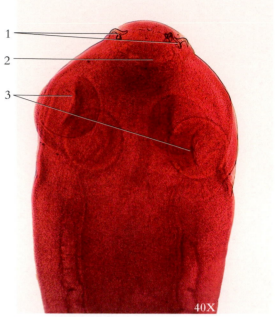

Figure 7.60 The scolex of *Taenia pisiformis*.
1. Hooks
2. Rostellum
3. Suckers

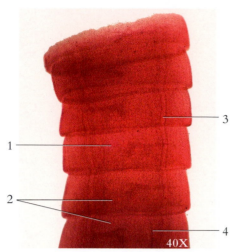

Figure 7.61 The immature proglottids
of *Taenia pisiformis*.
1. Early ovary
2. Early testes
3. Excretory canal
4. Immature vagina and ductus deferens

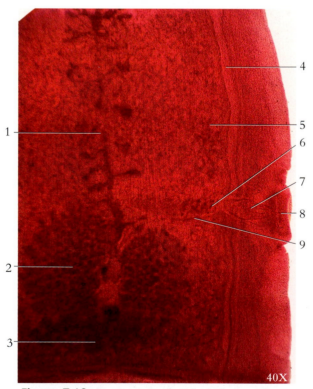

Figure 7.62 A mature proglottid of *Taenia pisiformis*.

1. Uterus	4. Excretory canal	7. Cirrus
2. Ovary	5. Testes	8. Genital pore
3. Yolk gland	6. Ductus deferens	9. Vagina

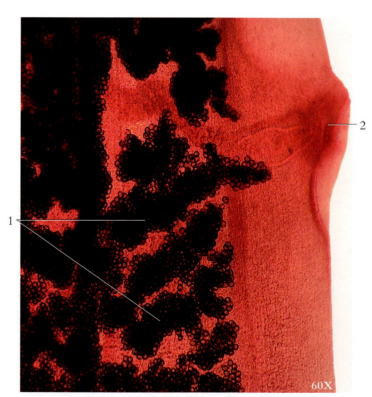

Figure 7.63 The ripe proglottid of *Taenia pisiformis*.
1. Zygotes in branched uterus   2. Genital pore

Phylum Mollusca

Table 7.4 Representatives of the Phylum Mollusca

Classes and Representative Kinds	Characteristics
Polyplacophora — chitons	Marine; shell of 8 dorsal plates; broad foot
Gastropoda — snails and slugs	Marine, freshwater, and terrestrial; coiled shell; prominent head with tentacles and eyes
Bivalvia — clams, oysters, and mussels	Marine, freshwater; body compressed between two hinged shells; hatchet-shaped foot
Cephalopoda — squids, octopi, and nautilus	Marine; excellent swimmers, predatory; foot separated into tentacles, which may contain suckers; well-developed eyes

Class Bivalvia — Foot, Clam

Class Monoplacophora — *Neopilina*

Class Polyplacophora — Girdle, *Chiton*

Class Gastropoda — Snail, Foot

Class Scaphopoda — Tooth shell, Foot

Class Cephalopoda — Head, Foot, Squid

Figure 7.64 Diagrams of specimens representing several classes of mollusks.

Figure 7.65 Chitons are easily recognized by their eight dorsal plates. (a) A dorsal view and (b) ventral view. (c) A ventral view of a chiton skeleton showing the eight dorsal plates.

1. Dorsal plates   2. Girdle   3. Mouth   4. Gill filaments   5. Ventral foot

Figure 7.66 A snail.
1. Shell
2. Foot
3. Occular tentacle
4. Head
5. Tentacle

Figure 7.67 The locomotion of the slug, class Gastropoda. This requires the production of mucus. Slugs differ from snails in that a shell is absent in the former.
1. Foot
2. Mucus
3. Mantle
4. Head
5. Occular tentacle
6. Sensory tentacle
7. Pneumostome

Figure 7.68 A snail radula, made up of small horny teeth of chitin, called denticles.

120X

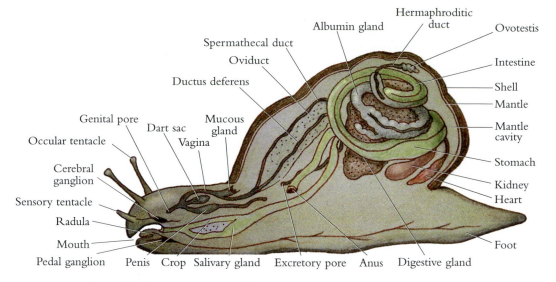

Figure 7.69 A diagram of pulmonate snail anatomy.

Figure 7.70 An view of a bivalve (clam) shell. Bivalves have two shells (valves) that are laterally compressed and dorsally hinged. (a) Dorsal view and (b) left valve.

1. Umbo   2. Hinge ligament   3. Growth lines

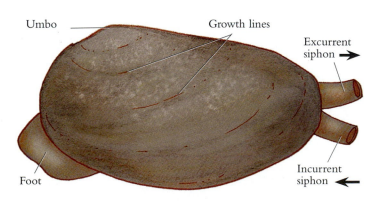

Figure 7.71 The surface anatomy of a freshwater clam, left valve.

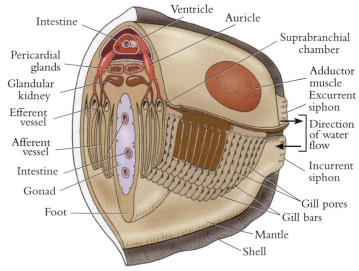

Figure 7.72 A diagram of the circulatory and respiratory systems of a freshwater clam.

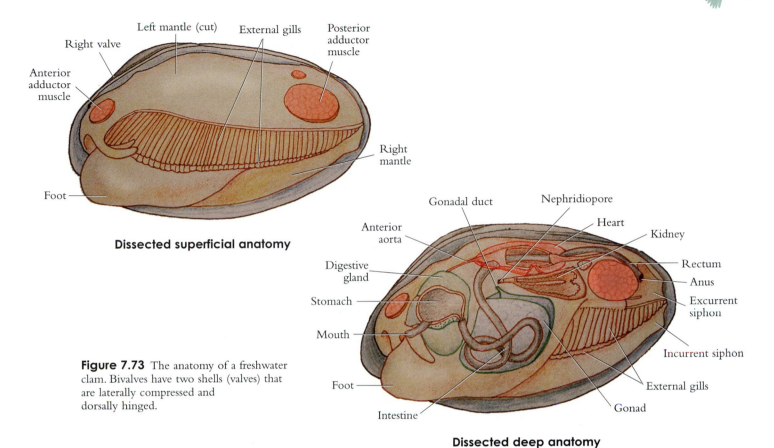

Figure 7.73 The anatomy of a freshwater clam. Bivalves have two shells (valves) that are laterally compressed and dorsally hinged.

Dissected superficial anatomy

Dissected deep anatomy

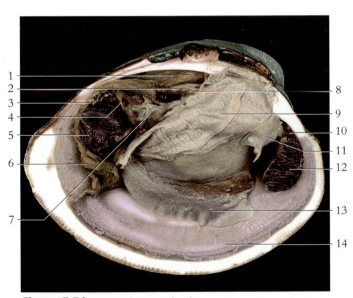

Figure 7.74 A lateral view of a clam.

1. Pericardium
2. Ventricle of heart
3. Anus
4. Posterior retractor muscle
5. Posterior adductor muscle
6. Excurrent siphon
7. Nephridium (kidney)
8. Atrium of heart
9. Gills
10. Anterior retractor muscle
11. Labial palps
12. Anterior adductor muscle
13. Foot
14. Mantle

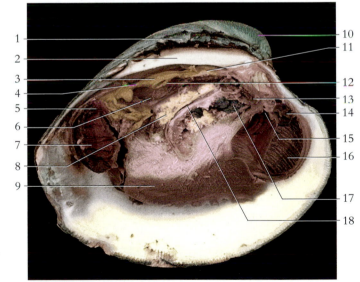

Figure 7.75 A lateral view of a clam, foot cut.

1. Hinge ligament
2. Hinge
3. Ventricle of heart
4. Posterior aorta
5. Posterior retractor muscle
6. Nephridium (kidney)
7. Posterior adductor muscle
8. Gonad
9. Foot
10. Umbo
11. Intestine
12. Opening between atrium and ventricle
13. Esophagus
14. Anterior retractor muscle
15. Mouth
16. Anterior adductor muscle
17. Digestive gland
18. Intestine

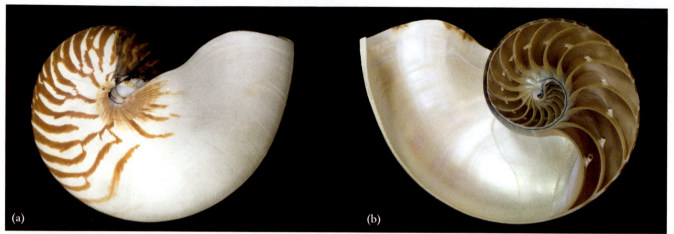

Figure 7.76 The *Nautilus*, a cephalopod, has gas-filled chambers within its shell, as seen in this cross-section of the shell (b). These chambers regulate buoyancy.

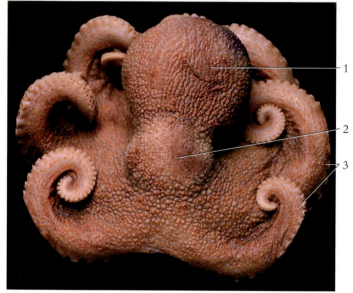

Figure 7.77 A dorsal view of an octopus collected in the Sea of Cortez, San Carlos, Mexico.

1. Mantle   2. Head   3. Arms

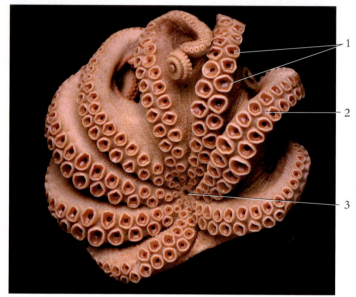

Figure 7.78 A ventral view of an octopus.

1. Suction cups   3. Mouth
2. Arm

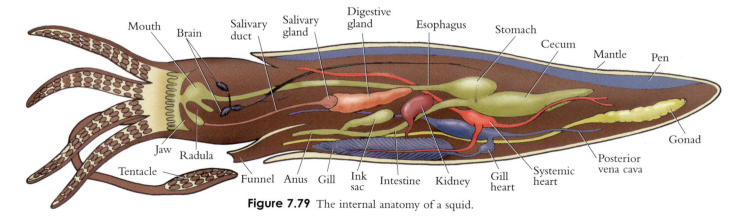

Figure 7.79 The internal anatomy of a squid.

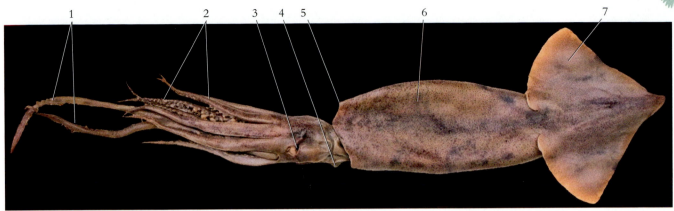

Figure 7.80 The external anatomy of the squid, *Loligo*.

1. Tentacles   2. Arms   3. Eye   4. Funnel   5. Collar   6. Mantle (body tube)   7. Fin

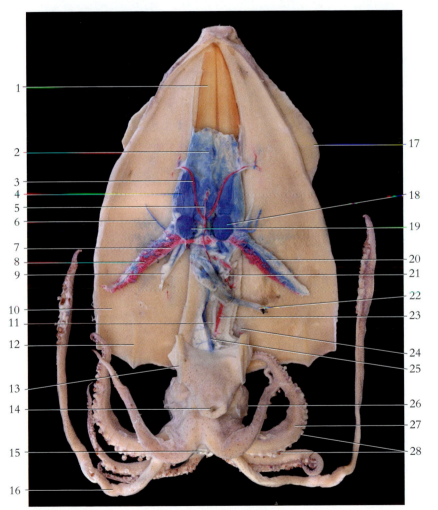

Figure 7.81 The internal anatomy of the squid.

1. Pen	11. Esophagus	20. Efferent branchial vein
2. Gonad	12. Articulating ridge	21. Ink sac
3. Lateral mantle artery	13. Articulating cartilage	22. Rectum
4. Posterior vena cava	14. Siphon	23. Cephalic aorta
5. Median mantle artery	15. Mouth	24. Stellate ganglion
6. Median mantle vein	16. Tentacle	25. Cephalic vena cava
7. Afferent branchial artery	17. Fin	26. Eye
8. Gill	18. Branchial heart	27. Arm
9. Genital opening	19. Systemic heart	28. Suckers
10. Mantle		

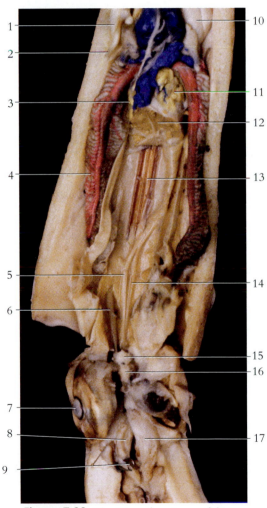

Figure 7.82 The internal anatomy of the squid including head region.

1. Spermatophoric duct	10. Stomach
2. Penis	11. Pancreas
3. Kidney	12. Digestive gland (cut)
4. Gill	13. Pen
5. Esophagus	14. Cephalic aorta
6. Pleural nerve	15. Visceral ganglion
7. Eye	16. Pedal ganglion
8. Radula	17. Buccal bulb
9. Beak	

Phylum Annelida

Table 7.5 Some Representatives of the Phylum Annelida

Classes and Representative Kinds	Characteristics
Polychaeta — tubeworms and sandworms	Mostly marine; segments with parapodia
Oligochaeta — earthworms	Freshwater and burrowing terrestrial forms; small setae; poorly developed head
Hirudinea — leeches	Freshwater; some are blood-sucking parasites and others are predators; lack setae; prominent muscular suckers at both ends

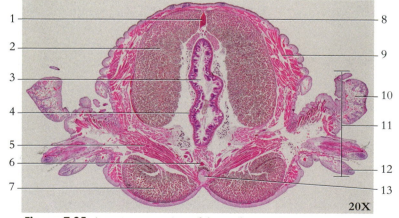

Figure 7.83 The sandworm, *Nereis* (scale in mm).

1. Parapodia   2. Mouth

Figure 7.84 The anterior end of the sandworm, *Nereis*. (a) A dorsal view and (b) a ventral view.

1. Palpi
2. Prostomium
3. Peristomial cirri
4. Peristome
5. Parapodia
6. Setae
7. Mouth
8. Everted pharynx

Figure 7.85 A transverse section of the sandworm, *Nereis*.

1. Dorsal blood vessel
2. Dorsal longitudinal muscle
3. Lumen of intestine
4. Intestine
5. Oblique muscle
6. Ventral blood vessel
7. Ventral longitudinal muscle
8. Integument
9. Circular muscle
10. Notopodium
11. Parapodium
12. Neuropodium
13. Ventral nerve cord

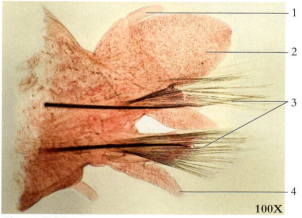

Figure 7.86 The parapodium of the sandworm, *Nereis*.

1. Dorsal cirrus
2. Notopodium
3. Setae
4. Neuropodium

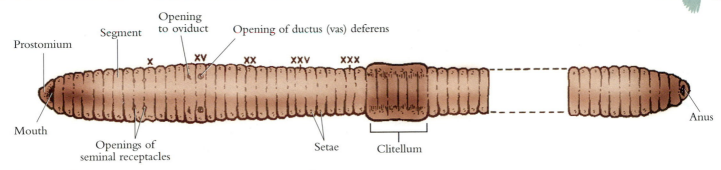

Figure 7.87 A ventral view diagram of the earthworm, *Lumbricus*.

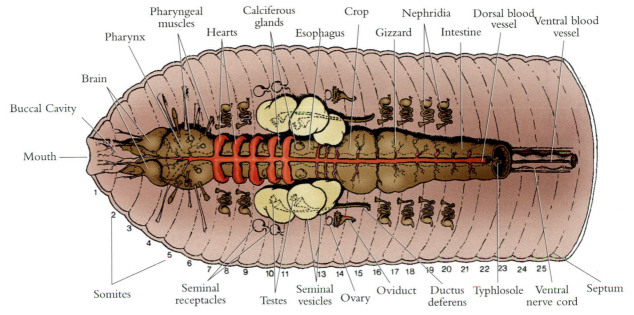

Figure 7.88 A dorsal view of the anterior end of the earthworm, *Lumbricus*.

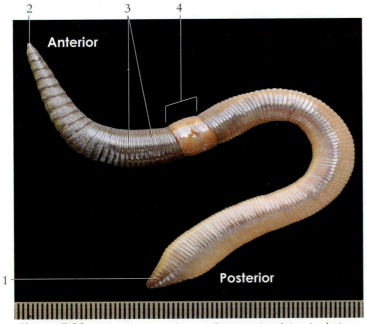

Figure 7.89 A dorsal view of an earthworm, *Lumbricus* (scale in mm).

1. Pygidium   3. Segments, or metameres
2. Prostomium   4. Clitellum

Figure 7.90 An anterior end of an earthworm, *Lumbricus* (scale in mm).

1. Prostomium   3. Setae   5. Opening of ductus
2. Mouth   4. Segment 10   (vas) deferens

Figure 7.91 Earthworm cocoons (scale in mm).

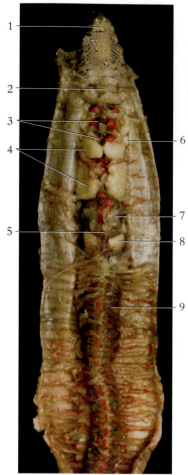

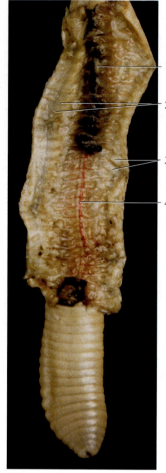

Figure 7.92 The internal anatomy of the anterior end of an earthworm, *Lumbricus.*

1. Brain
2. Pharynx
3. Hearts
4. Seminal vesicles
5. Dorsal blood vessel
6. Seminal receptacles
7. Crop
8. Gizzard
9. Intestine

Figure 7.93 The internal anatomy of the posterior end of an earthworm with part of the intestine removed.

1. Intestine
2. Septae
3. Nephridia
4. Ventral blood vessel

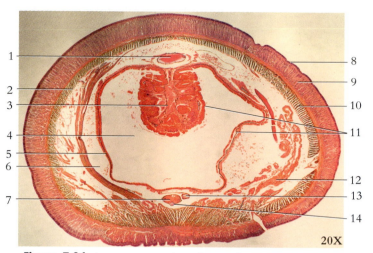

Figure 7.94 A transverse section of an earthworm posterior to the clitellum.

1. Dorsal blood vessel
2. Peritoneum
3. Typhlosole
4. Lumen of intestine
5. Intestine
6. Coelom
7. Ventral nerve cord
8. Epidermis
9. Circular muscles
10. Longitudinal muscles
11. Intestinal epithelium
12. Nephridium
13. Ventral blood vessel
14. Subneural blood vessel

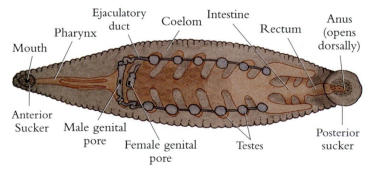

Figure 7.95 A diagram of a leech.

Mouth
Pharynx
Ejaculatory duct
Coelom
Intestine
Rectum
Anus (opens dorsally)
Anterior Sucker
Male genital pore
Female genital pore
Testes
Posterior sucker

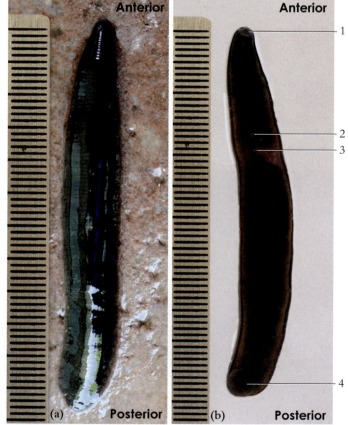

Figure 7.96 (a) A dorsal view of a leech and (b) a ventral view of a leech. Leeches are more specialized than other annelids. They have lost their setae and developed suckers for attachment while sucking blood (scale in mm).

1. Anterior sucker
2. Male genital pore
3. Female genital pore
4. Posterior sucker

Phylum Nematoda

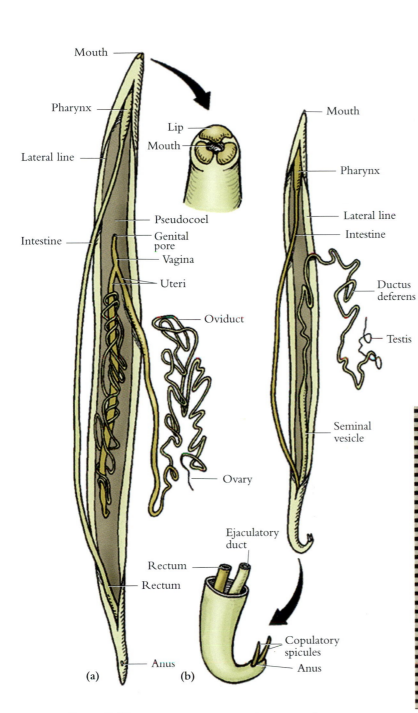

Figure 7.97 A diagram of the internal anatomy of (a) a female and (b) a male *Ascaris*.

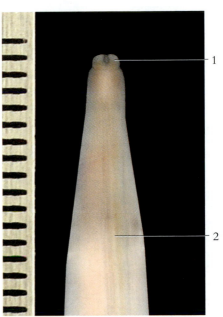

Figure 7.98 The head end of a male *Ascaris* (scale in mm).

1. Lip
2. Lateral line

Figure 7.99 The posterior end of (a) a female and (b) a male *Ascaris* (scale in mm).

1. Copulatory spicules
2. Ejaculatory duct

Figure 7.100 The internal anatomy of a male *Ascaris* (scale in mm).

1. Intestine
2. Lateral line
3. Ductus deferens
4. Testes
5. Seminal vesicle

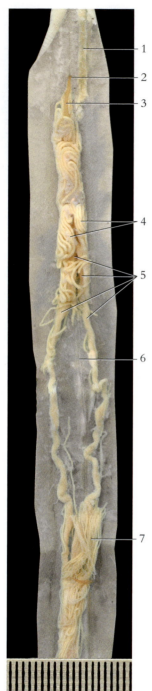

Figure 7.101 The internal anatomy of a female *Ascaris* (scale in mm).

1. Intestine
2. Genital pore
3. Vagina
4. Oviducts
5. Uteri (Y-shaped)
6. Lateral line
7. Ovary

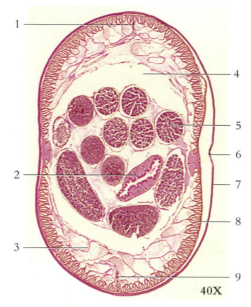

40X

Figure 7.102 A transverse sections of a male *Ascaris*.

1. Dorsal nerve cord
2. Intestine
3. Longitudinal muscle cell body
4. Pseudocoel
5. Testis
6. Lateral line
7. Cuticle
8. Contractile sheath of muscle cell
9. Ventral nerve cord

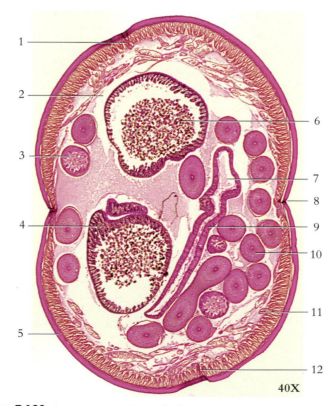

40X

Figure 7.103 A transverse sections of a female *Ascaris*.

1. Dorsal nerve cord
2. Pseudocoel
3. Oviduct
4. Uterus
5. Cuticle
6. Eggs
7. Lumen of intestine
8. Lateral line
9. Intestine
10. Ovary
11. Longitudinal muscles
12. Ventral nerve cord

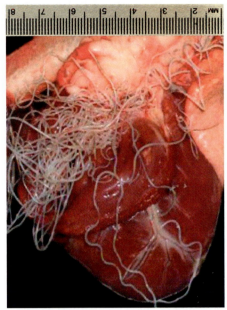

Figure 7.104 A dog heart infested with heartworm, *Dirofilaria immitis* (scale in mm).

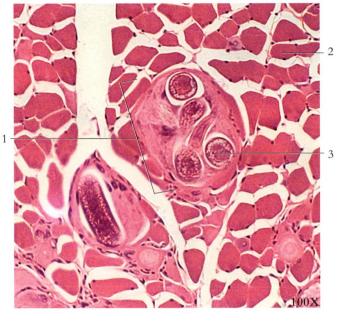

Figure 7.105 A photomicrograph of *Trichinella spiralis* encysted in muscle.
1. Cyst   3. Larva
2. Muscle

Phylum Rotifera

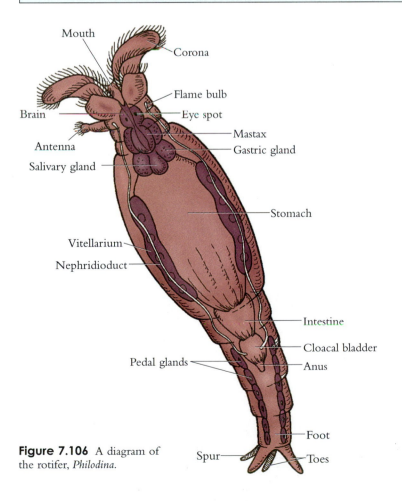

Figure 7.106 A diagram of the rotifer, *Philodina*.

Mouth
Corona
Flame bulb
Eye spot
Brain
Mastax
Antenna
Gastric gland
Salivary gland
Stomach
Vitellarium
Nephridioduct
Intestine
Cloacal bladder
Pedal glands
Anus
Foot
Spur
Toes

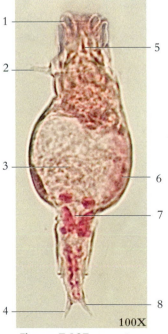

Figure 7.107 A rotifer.
1. Corona   5. Mastax
2. Antenna   6. Vitellarium
3. Stomach   7. Intestine
4. Spur   8. Toe

Phylum Arthropoda

Table 7.6 Representatives of the Phylum Arthropoda

Subphyla: Classes and Representative Kinds	Characteristics
Chelicerata: Merostomata — horseshoe crab	Cephalothorax and abdomen; specialized front appendages into chelicerae; lack antennae and mandibles
Chelicerata: Arachnida — spiders, mites, ticks, and scorpions	Cephalothorax and abdomen; four pairs of legs; book lungs or trachea; lack antennae and mandibles
Crustacea: Malacostraca — lobsters, crabs, and shrimp	Cephalothorax and abdomen; two pairs of antennae; pair of mandibles and two pairs of maxillae; biramous appendages; gills
Crustacea: Maxillopoda — copepods and barnacles	Cephalothorax and abdomen; freshwater and marine; basic plan of five head and ten trunk segments; terminal telson; up to six pairs of appendages
Hexapoda: Insecta — beetles, butterflies, and ants	Head, thorax, and abdomen; three pairs of legs; well-developed mouth parts; usually two pair of wings; trachea
Myriapoda: Chilopoda — centipedes	Head with segmented trunk; one pair of legs per segment; trachea; one pair of antennae
Myriapoda: Diplopoda — millipedes	Head with segmented trunk; usually two pair of legs per segment; trachea

(a) (b)

Figure 7.108 Trilobites are extinct arthropods from the Cambrian and Ordovician periods. (a) *Modicia typicalis*, and (b) *Dicranurus elegans*.

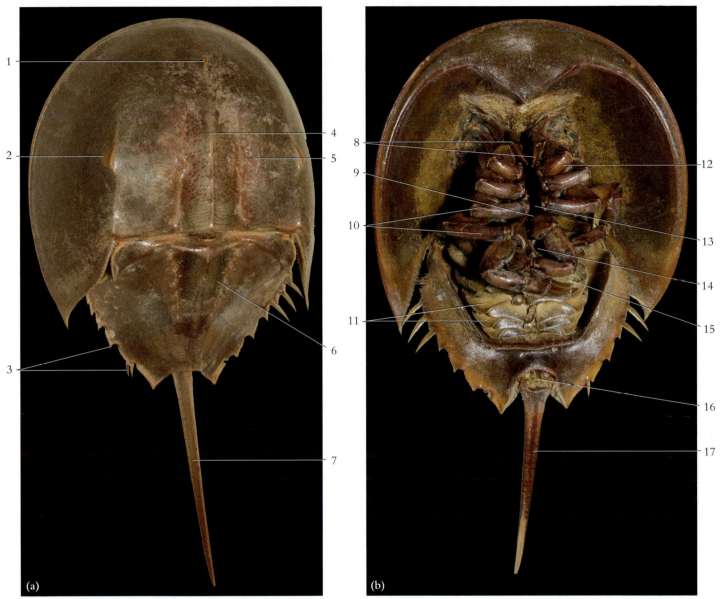

Figure 7.109 (a) A dorsal view and (b) a ventral view of the horseshoe crab, *Limulus*. This animal is commonly found in shallow waters along the Atlantic coast from Canada to Mexico.

1. Simple eye	6. Abdomen (opisthosoma)	11. Book gills	16. Anus
2. Compound eye	7. Telson	12. Pedipalp	17. Telson
3. Abdominal spines	8. Chelicerae	13. Mouth	
4. Anterior spine	9. Gnathobase	14. Chilarium	
5. Cephalothorax (prosoma)	10. Chelate legs	15. Genital operculum	

Figure 7.110 A beach spider, *Argiope appensa*, an introduced species to Hawaii.

Figure 7.111 A garden spider in the process of spinning a web.
1. Spinnerets

Figure 7.112 A tick, within the family Ixodidae. It is a specialized parasitic arthropod (scale in mm).

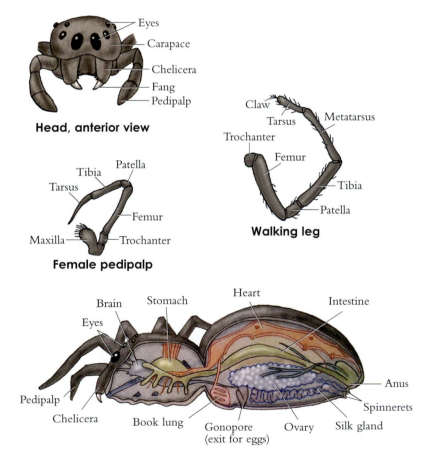

Eyes
Carapace
Chelicera
Fang
Pedipalp

Head, anterior view

Tibia
Patella
Tarsus
Femur
Maxilla
Trochanter

Female pedipalp

Claw
Tarsus
Metatarsus
Trochanter
Femur
Tibia
Patella

Walking leg

Brain
Stomach
Heart
Intestine
Eyes
Pedipalp
Chelicera
Book lung
Gonopore (exit for eggs)
Ovary
Silk gland
Spinnerets
Anus

Figure 7.113 A diagram of the anatomy of a spider.

Figure 7.114 A rose-haired tarantula, *Grammostola cola*.
1. Opisthosoma   2. Pedipalp   3. Prosoma

Figure 7.115 An Arizona hairy scorpion, *Hadrurus arizonensis*. Scorpions are found most commonly in tropical and subtropical regions. However, there are a number of species found in arid and temperate zones.

1. Pedipalp
2. Cephalothorax
3. Walking legs
4. Preabdomen
5. Stinging apparatus
6. Postabdomen (tail)

Figure 7.116 Some ticks attached and feeding on a savannah monitor, a large African lizard.

1. Ticks
2. Scales of monitor

Figure 7.117 A peppermint shrimp, *Lysmata wurdemanni*.

Figure 7.118 A ghost crab, *Ocypode ceratophthalmus*.

Figure 7.119 A hermit crab, *Coenobita clypeatus*.

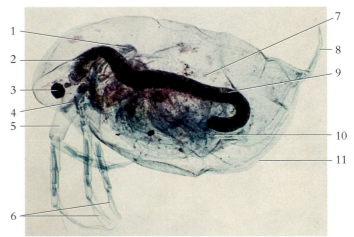

Figure 7.120 The water flea, *Daphnia*, a common microscopic crustacean.

1. Heart
2. Midgut
3. Compound eye
4. Mouth
5. 2nd antenna
6. Setae
7. Brood chamber
8. Abdominal seta
9. Hindgut
10. Anus
11. Carapace

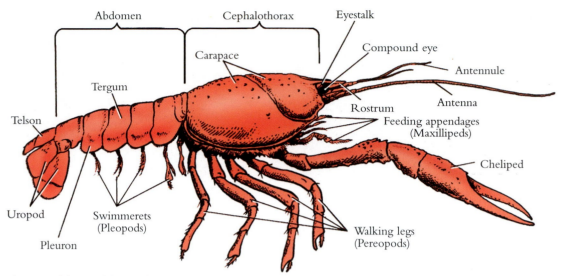

Figure 7.121 A diagram of the crayfish, *Cambarus*.

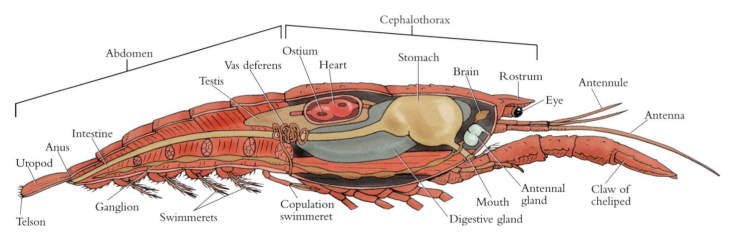

Figure 7.122 The anatomy of a crayfish; sagittal section of an adult male.

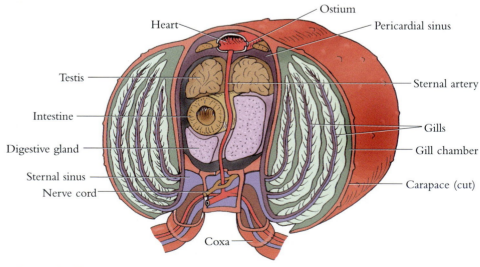

Figure 7.123 The anatomy of a crayfish; transverse section of an adult male.

Figure 7.125 A lateral view of the crayfish.

1. Carapace
2. Abdomen
3. Uropod
4. Swimmeret
   (pleopod)
5. Rostrum
6. Compound eye
7. Maxilliped
8. Cheliped
9. Walking legs

Figure 7.124 A dorsal view of the crayfish.

1. Cheliped
2. Walking legs
3. Carapace
4. Abdomen
5. Telson
6. Uropod
7. Antenna
8. Antennule
9. Rostrum
10. Compound eye
11. Cephalothorax
12. Tergum

Figure 7.126 A ventral view of the oral region of the crayfish.

1. Third maxilliped
2. Second maxilla
3. First maxilliped
4. Green gland duct
5. Mandible

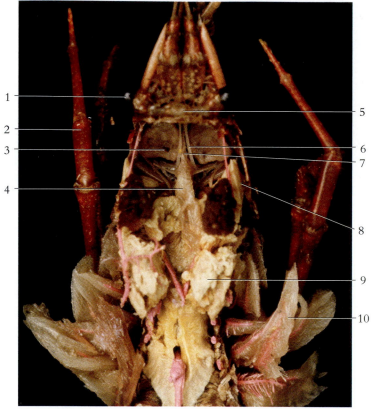

Figure 7.127 A dorsal view of the oral region of the crayfish.

1. Compound eye
2. Walking leg
3. Green gland
4. Cardiac chamber
   of stomach
5. Brain
6. Circumesophageal
   connection (of ventral nerve cord)
7. Esophagus
8. Region of gastric mill
9. Digestive gland
10. Gill

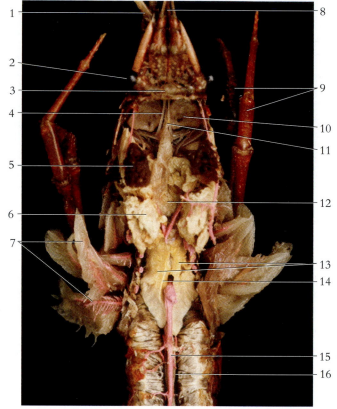

Figure 7.128 A dorsal view of the anatomy of a crayfish.

1. Antenna
2. Compound eye
3. Brain
4. Circumesophageal
   connection (of
   ventral nerve cord)
5. Mandibular muscle
6. Digestive gland
7. Gills
8. Antennules
9. Walking legs
10. Green gland
11. Esophagus
12. Pyloric stomach
13. Testis
14. Ductus
    deferens
15. Aorta
16. Intestine

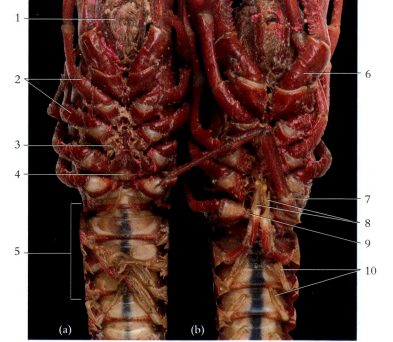

Figure 7.129 A ventral views of (a) a female
crayfish and (b) a male crayfish. The first pair of
swimmerets are greatly enlarged in the male for
the depositing of sperm in the female's seminal
receptacle.

1. Third maxilliped
2. Walking legs
3. Disk covering oviduct
4. Seminal receptacle
5. Abdomen
6. Base of cheliped
7. Base of last walking leg
8. Copulatory swimmerets (pleopods)
9. Sperm ducts (genital pores)
10. Swimmerets (pleopods)

Figure 7.130 An angular winged katydid, *Microcentrum retinerve*.

Figure 7.131 A red milkweed beetle, *Tetraopes tetraophthalmus*.

Figure 7.132 A flame skimmer dragonfly, *Libellula saturata*.

Figure 7.133 A seven-spot ladybird beetle, *Coccinella septempunctata*.

Figure 7.134 A spicebush swallowtail butterfly, *Papilio troilus*.

Figure 7.135 A field cricket, *Gryllus firmus*.

Figure 7.136 The developmental stages of the common honeybee: (a) larval stage, (b) pupa, and (c) adult (scale in mm).

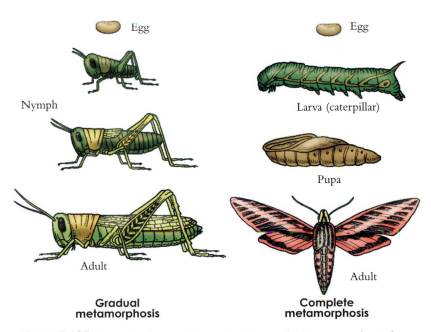

Figure 7.137 Insect development. In gradual (incomplete) metamorphosis the young resemble the adults, but they are smaller and have different body proportions. In complete metamorphosis, the larvae look different from the adult and generally have different food requirements.

Figure 7.138 The developmental stages of the Monarch butterfly, *Danaus plexippus*, include (a) egg, (b) larval stage, (c) chrysalis, and (d) adult.

Figure 7.139 A common gray cricket molting. All arthropods must periodically shed their exoskeleton in order to grow. This process is called molting, or ecdysis.

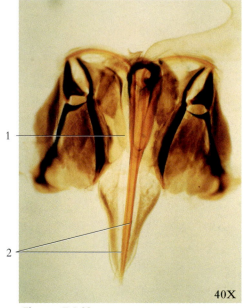

40X

Figure 7.140 A honeybee stinger. The two darts contain barbs on the tips, which point upward, making it difficult to remove a stinger from a wound.

1. Sheath   2. Darts

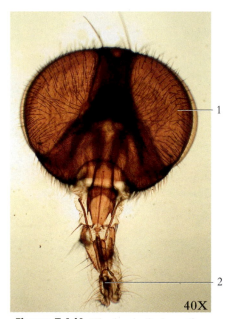

40X

Figure 7.141 The head of a housefly, an example of a sponging type of mouthpart in insects. Notice the large lobes at the apex of the labium, which function in lapping up liquids.

1. Compound eye   2. Labium

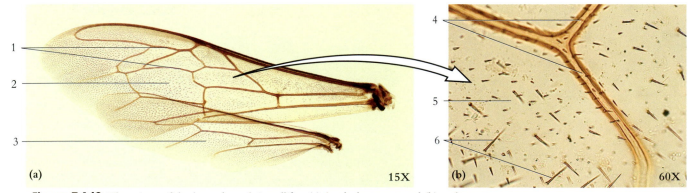

(a)   15X   (b)   60X

Figure 7.142 The wings of the honeybee, *Apis mellifera*. (a) A whole mount and (b) a close-up.

1. Cross veins   3. Hindwing   5. Transparent wing film
2. Forewing   4. Cross veins   6. Hairs

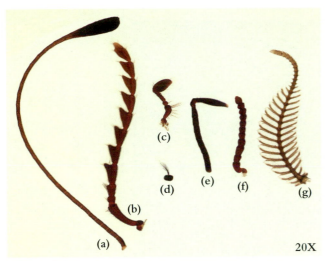

Figure 7.143 Some common insect antennae. (a) Clavate - butterflies, (b) serrate - click beetles, (c) lamellate - scarab beetles, (d) aristate - house flies, (e) geniculate - weevils (f) moniliform - termites, and (g) plumose - moths.

Figure 7.144 The plumose antennae of the Ailanthus silkmoth, *Samia cynthia*.

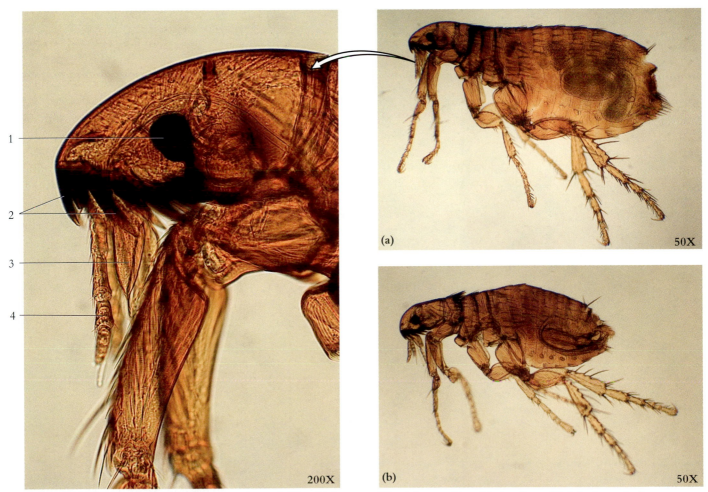

Figure 7.145 The mouthparts of the flea, *Ctenocephalide*, which are specialized for parasitism. Notice the oral bristles beneath the mouth that aid the flea in penetrating between hairs to feed on the blood of mammals. (a) Female flea and (b) male flea.

1. Eye   2. Oral bristles   3. Maxilla   4. Maxillary palp

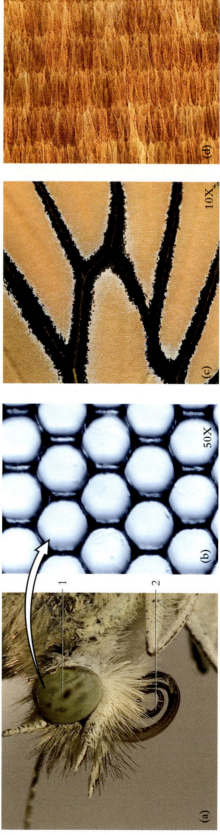

Figure 7.146 (a) A lateral view of the head of a butterfly. The most obvious structures on the head of a butterfly are compound eyes and the curled tongue for siphoning nectar from flowers. (b) A magnified view of the compound eye. (c) A close-up view of the wing scales and (d) a magnified view of the wing scales.

1. Compound eye   2. Tongue

Figure 7.148 A preserved specimen of a grasshopper, order Orthoptera.

1. Mesothorax   7. Maxilla   13. Mesothorax
2. Pronotum   8. Antenna   14. Spiracle
3. Compound eye   9. Claw   15. Abdomen
4. Vertex   10. Wing   16. Trochanter
5. Gena   11. Femur   17. Tarsus
6. Frons   12. Tibia

Figure 7.147 The surface anatomy of a grasshopper, order Orthoptera.

1. Antenna   7. Labrum   13. Tergum
2. Compound eye   8. Labial palp   14. Spiracle
3. Ocelli   9. Femur   15. Cercus
4. Prothorax   10. Metathorax   16. Ovipositor
5. Mesothorax   11. Tibia   17. Tarsus
6. Tympanum   12. Forewing   18. Sternum

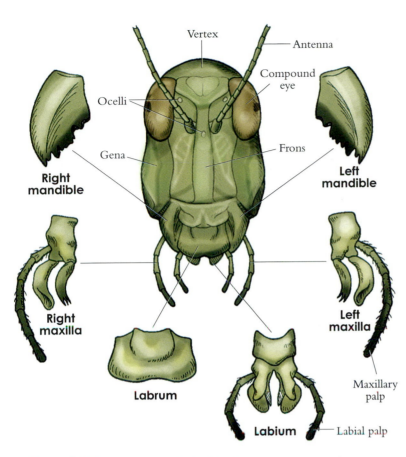

Figure 7.149 A diagram of the head and mouthparts of a grasshopper.

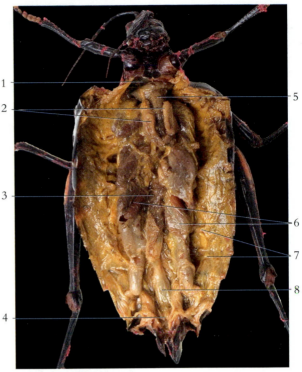

Figure 7.150 A ventral view showing the internal anatomy of a dissected grasshopper.

1. Esophagus   6. Ovaries
2. Gastric caecum   7. Trachea
3. Stomach   8. Intestine
4. Rectum
5. Crop

Figure 7.151 A garden centipede, *Lithobius* (scale in mm).

Figure 7.152 A Vietnamese centipede, *Scolopendra subspinipes* (scale in mm).

Figure 7.153 An African millipede, *Scaphiostreptus* (scale in mm).

Figure 7.154 A garden millipede, *Chordeumatida* (scale in mm).

Phylum Echinodermata

Table 7.7 Representatives of the Phylum Echinodermata

Classes and Representative Kinds	Characteristics
Asteroidea — sea stars (starfish)	Pentaradial symmetry; appendages arranged around a central disk containing the mouth; tube feet with suckers
Echinoidea — sea urchins and sand dollars	Pentaradial symmetry; disk-shaped with no arms; compact skeleton; movable spines; tube feet with suckers
Ophiuroidea — brittle stars	Pentaradial symmetry; appendages sharply marked off from central disk; tube feet without suckers
Holothuroidea — sea cucumbers	Pentaradial symmetry; cucumber-shaped with no arms; spines absent; tube feet with tentacles and suckers
Crinoidea — sea lilies and feather stars	Pentaradial symmetry; sessile during much of life cycle; calyx supported by elongated stalk in some

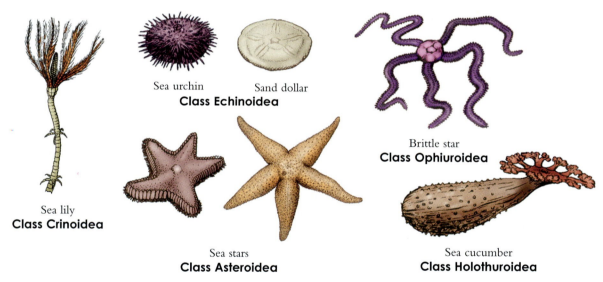

Sea urchin   Sand dollar
Class Echinoidea

Brittle star
Class Ophiuroidea

Sea lily
Class Crinoidea

Sea stars
Class Asteroidea

Sea cucumber
Class Holothuroidea

Figure 7.155 A diagram of specimens representing each class of echinoderms.

Figure 7.156 A group of sea stars (starfish), *Asterias*, in a tide pool in Oregon.

Figure 7.157 A green sea urchin, *Strongylocentrotus droebachiensis*.

Figure 7.158 A California sea cucumber, *Parastichopus californicus*.

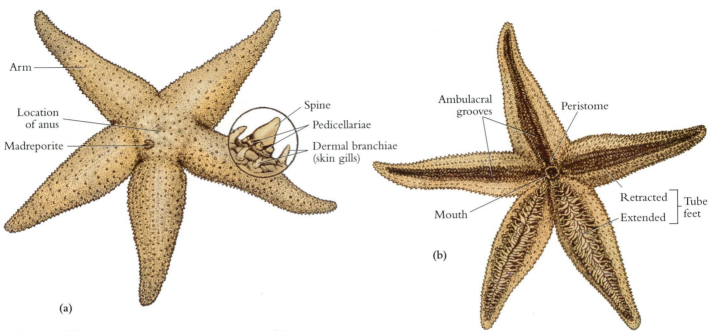

Figure 7.159 A diagram of the external anatomy of the sea star (starfish), *Asterias*. (a) An aboral (dorsal) view and (b) an oral (ventral) view.

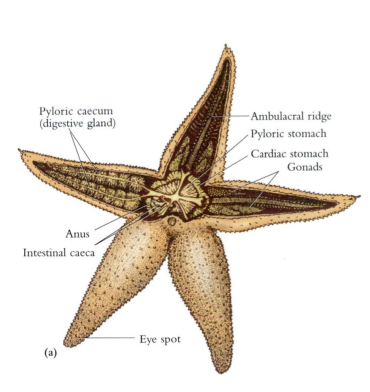

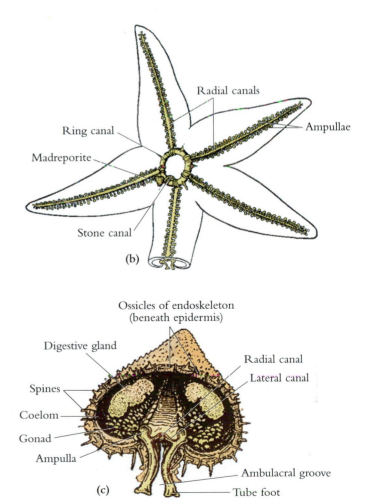

Figure 7.160 A diagram of the internal anatomy of the sea star from aboral side. (a) The digestive and reproductive organs, (b) the water vascular system, and (c) a transverse section through an arm.

Figure 7.161 An oral view of a sea star (a) showing the cardiac stomach extended through mouth and (b) after retracting stomach.
1. Cardiac stomach

Figure 7.162 An aboral view of the internal anatomy of a sea star.
1. Ambulacral ridge   3. Spines   5. Pyloric caecum
2. Gonad   4. Ring canal   (digestive gland)

Figure 7.163 A magnified aboral view of the internal anatomy of a sea star.
1. Ambulacral ridge   6. Pyloric duct   10. Pyloric
2. Madreporite   7. Pyloric caecum   stomach
3. Stone canal   (digestive gland)   11. Spines
4. Ampullae   8. Gonad
5. Polian vesicle   9. Anus

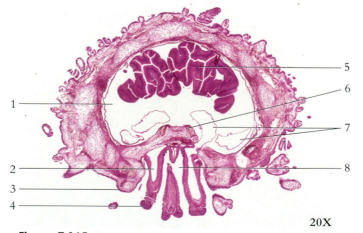

Figure 7.164 An oral view of a sea star.
1. Tube feet   4. Ambulacral groove
2. Peristome   5. Oral spines
3. Mouth

Figure 7.165 A transverse section through the arm of a sea star.
1. Coelom   4. Sucker   7. Gonad
2. Tube foot   5. Pyloric caecum   8. Ambulacral groove
3. Epidermis   6. Radial canal

20X

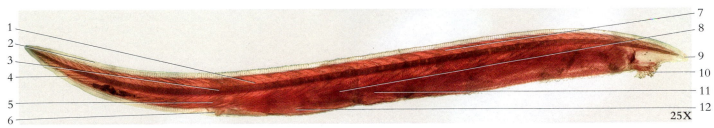

Figure 7.177 A whole mount of *Amphioxus*.

1. Dorsal nerve cord
2. Caudal fin
3. Notochord
4. Anus
5. Intestine
6. Atriopore
7. Myomeres
8. Esophagus
9. Rostrum
10. Oral cirri
11. Hepatic caecum
12. Atrium

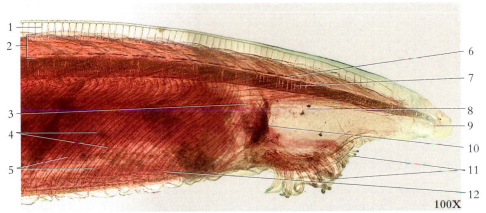

Figure 7.178 An anterior view of the anatomy of *Amphioxus*.

1. Fin rays
2. Myomeres
3. Velum
4. Gill slits
5. Gill bars
6. Dorsal nerve cord
7. Notochord
8. Oral hood
9. Rostrum
10. Wheel organ
11. Oral cirri
12. Pharynx

Figure 7.179 A posterior view of the anatomy of *Amphioxus*.

1. Caudal fin
2. Fin rays
3. Notochord
4. Anus
5. Intestine
6. Myomeres
7. Midgut
8. Atrium
9. Atriopore

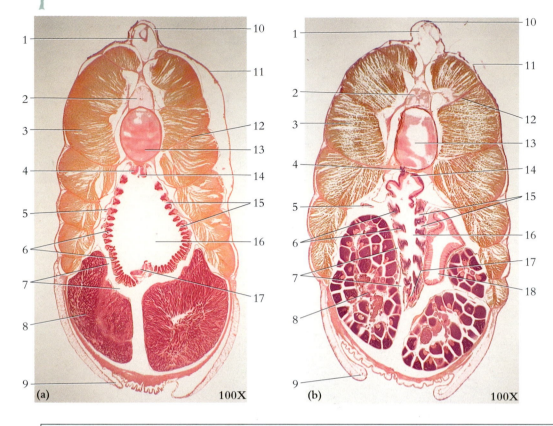

Figure 7.180 A transverse section through the pharyngeal region of (a) a male, and (b) a female *Amphioxus*.

1. Fin ray
2. Dorsal nerve cord
3. Myomere
4. Dorsal aorta
5. Nephridium
6. Gill bars
7. Atrium
8. Testis (male)
   Ovary (female)
9. Metapleural fold
10. Dorsal fin
11. Epidermis
12. Myoseptum
13. Notochord
14. Epibranchial groove
15. Gill slits
16. Pharynx
17. Endostyle (hypobranchial groove)
18. Hepatic caecum (liver)

(a) 100X
(b) 100X

Subphylum Vertebrata   Class Cephalaspidomorphi

Figure 7.181 A Pacific lamprey, *Lampetra tridentata*.

Figure 7.182 A dorsal view of the external anatomy of a marine lamprey, *Petromyzon marinus*.

1. Head
2. Nostril
3. Pineal body
4. Caudal fin
5. Posterior dorsal fin
6. Trunk
7. Anterior dorsal fin

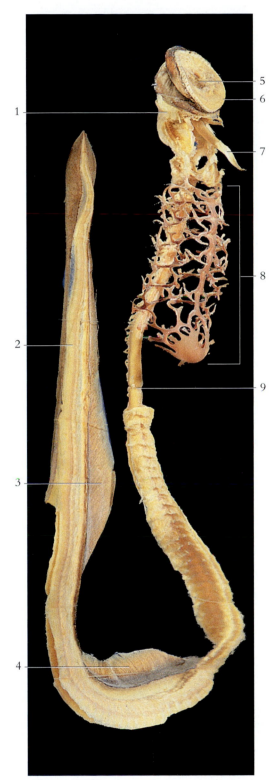

Figure 7.183 The cartilaginous skeleton of a marine lamprey.

1. Cranium
2. Caudal fin
3. Posterior dorsal fin
4. Anterior dorsal fin
5. Buccal cavity

6. Annular cartilage
7. Lingual cartilage
8. Branchial basket
9. Notochord

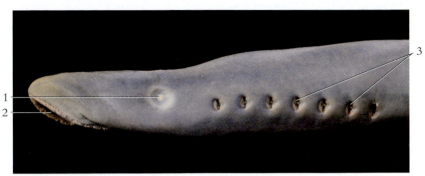

Figure 7.184 A lateral view of the anterior anatomy of a marine lamprey.

1. Eye
2. Buccal funnel

3. External gill slits

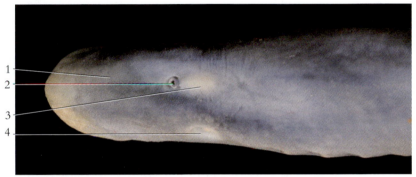

Figure 7.185 A dorsal view of the anterior anatomy of a marine lamprey.

1. Head
2. Nostril

3. Pineal body
2. Eye

Figure 7.186 The oral region of a marine lamprey.

1. Horny teeth
2. Buccal papillae

3. Mouth

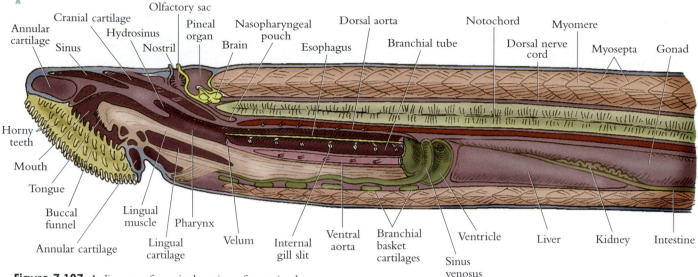

Figure 7.187 A diagram of a sagittal section of a marine lamprey.

Figure 7.188 A sagittal section through the anterior region of a lamprey.

1. Pineal organ	5. Mouth	9. Buccal muscle	13. Dorsal aorta
2. Nostril	6. Annular cartilage	10. Myomeres	14. Atrium
3. Brain	7. Lingual cartilage	11. Dorsal nerve cord	15. Ventricle
4. Pharynx	8. Internal gill slit	12. Notochord	16. Liver

Figure 7.189 A transverse section through the head at the level of the eyes of a lamprey.

1. Pineal organ
2. Brain
3. Lens of eye
4. Retina of eye
5. Lingual cartilage
6. Myomere
7. Cranial cartilage
8. Nasopharyngeal pouch
9. Pharynx
10. Pharyngeal gland

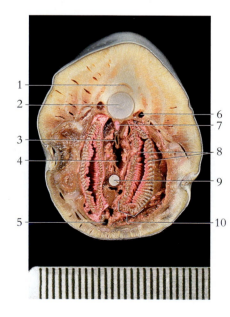

Figure 7.190 A transverse section through the body at the level of the fourth gill slit of a lamprey.

1. Spinal cord
2. Notochord
3. Esophagus
4. Respiratory tube
5. Ventral jugular vein
6. Anterior cardinal vein
7. Dorsal aorta
8. Gill filaments
9. Ventral aorta
10. Branchial pouch

Figure 7.191 A black tip reef shark, *Carcharhinus melanopterus*, class Chondrichthyes.

Figure 7.192 A gray reef shark, *Carcharhinus amblyrhynchos*, class Chondrichthyes..

Figure 7.193 A gray smoothhound shark, *Mustelus californicus*, class Chondrichthyes.

Figure 7.194 A guitarfish, *Rhina ancylostoma*, class Chondrichthyes.

Figure 7.195 A blue-spotted stingray, *Taeniura lymma*, class Chondrichthyes.

Figure 7.196 A nurse shark, *Ginglymostoma cirratum*, class Chondrichthyes.

Figure 7.197 The coelacanth, *Latimeria chalumnae*, in the class Sarcopterygii. This lobe-fin fish was once thought to be extinct.

Eye   Lateral line   Dorsal fin   Caudal fin

Nostril
Premaxilla

Dentary

Maxilla   Operculum   Pectoral fin   Pelvic fin   Anal fin

Figure 7.198 The external structures of a grouper, *Mycteroperca bonaci*, class Actinopterygii.

Figure 7.199 A red discus fish, *Symphysodon aequifasciatus*, class Actinopterygii.

Figure 7.200 A frogfish, *Antennarius* sp., class Actinopterygii.

Figure 7.201 A jack mackerel, *Trachurus declivis*, class Actinopterygii.

Figure 7.202 An Eastern newt, *Notophthalmus viridescens,* class Amphibia.

Figure 7.203 A tiger salamander, *Ambystoma tigrinum,* class Amphibia.

Figure 7.204 An American bullfrog, *Rana catesbeiana,* class Amphibia.

Figure 7.205 A poison-dart frog, *Dendrobates* sp., class Amphibia.

Figure 7.206 An African clawed frog, *Xenopus laevis,* class Amphibia.

Figure 7.207 A Woodhouse's toad, *Bufo woodhousii,* class Amphibia.

Figure 7.208 The surface anatomy and body regions of the leopard frog, *Rana pipiens,* class Amphibia.

1. Ankle	4. Eyes	7. Brachium
2. Knee	5. Nostril	8. Antebrachium
3. Foot	6. Tympanic membrane	9. Digits

Figure 7.209 A white-lipped tree frog, *Litoria infrafrenata,* class Amphibia. (a) The frog is crouched on a person's fingers. (b) The adhesive toe disks can be seen in a ventral view.

Figure 7.210 A horned lizard, *Phrynosoma platyrhinos.*, class Sauropsida (= Reptilia).

Figure 7.211 A western coachwhip snake *Masticophis flagellum*, class Sauropsida (= Reptilia).

Figure 7.212 A spiney soft shell turtle, *Apalone spinifera*, class Sauropsida (= Reptilia).

Figure 7.213 An Aldabra giant tortoise, *Dipsochelys dussumieri*, class Sauropsida (= Reptilia).

Figure 7.214 An American alligator, *Alligator mississippiensis*, class Sauropsida (= Reptilia).

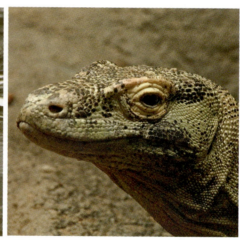

Figure 7.215 A Komodo dragon, *Varanus komodoensis*, class Sauropsida (= Reptilia).

Figure 7.216 A veiled chameleon, *Chamaeleo calyptratus*, class Sauropsida (= Reptilia).

Figure 7.217 A checkered whiptail lizard, *Cnemidophorus grahamii*, class Sauropsida (= Reptilia).

Figure 7.218 A Great Basin rattlesnake, *Crotalus viridis lutosis*, class Sauropsida (= Reptilia).

Figure 7.219 A chukar, *Alectoris chukar,* class Aves.

Figure 7.220 A robin, *Turdus migratorius,* class Aves.

Figure 7.221 A turkey vulture, *Cathartes aura,* class Aves.

Figure 7.222 A great horned owl, *Bubo virginianus,* class Aves.

Figure 7.223 An anhinga, *Anhinga anhinga,* class Aves.

Figure 7.224 The Magellanic penguin, *Spheniscus magellanicus,* class Aves.

Figure 7.225 A California quail, *Callipepla californica,* class Aves.

Figure 7.226 A great egret, *Ardea alba,* class Aves.

Figure 7.227 An American kestrel, *Falco sparverius,* class Aves.

Figure 7.228 A bottlenose dolphin, *Tursiops truncatus*, class Mammailia.

Figure 7.229 A California sea lion, *Zalophus californianus*, class Mammailia.

Figure 7.230 A nine-banded armadillo, *Dasypus novemcinctus*, class Mammailia.

Figure 7.231 A koala, *Phascolarctos cinereus*, class Mammailia.

Figure 7.232 A grizzly bear, *Ursus arctos horribilis*, class Mammailia.

Figure 7.233 A gray squirrel, *Sciurus carolinensis*, class Mammailia.

Figure 7.234 A mule deer, *Odocoileus hemionus*, class Mammailia.

Figure 7.235 A Malaysian fruit bat, *Pteropus hypomelanus*, class Mammailia.

Figure 7.236 An orangutan, *Pongo pygmaeus*, class Mammailia.

An understanding of the structure of a vertebrate organism is requisite to learning about physiological mechanisms and about how the animal functions in its environment. The selective pressures that determine evolutionary change frequently have an influence on anatomical structures. Studying dissected specimens, therefore, provides phylogenetic information about how groups of organisms are related.

Some biology laboratories have the resources to provide students with opportunities for doing selected vertebrate dissections. For these students, the photographs contained in this chapter will be a valuable source for identification of structures on your specimens as they are dissected and studied. If dissection specimens are not available, the excellent photographs of carefully dissected prepared specimens presented in this chapter will be an adequate substitute. Care has gone into the preparation of these specimens to depict and identify the principal body structures from representative specimens of each of the classes of vertebrates. Selected human cadaver dissections are shown in photographs contained in chapter 9. As the anatomy of vertebrate specimens is studied in this chapter, observe the photographs of human dissections in the next chapter and note the similarities of body structure, particularly to those of another mammal.

Class Chondrichthyes

Figure 8.1 A lateral view of the leopard shark, *Triakis semifasciata*.

1. Spiracle
2. Lateral line
3. Anterior dorsal fin
4. Posterior dorsal fin
5. Caudal fin (heterocercal tail)
6. Eye
7. Gill slits
8. Pectoral fin
9. Pelvic fin
10. Anal fin

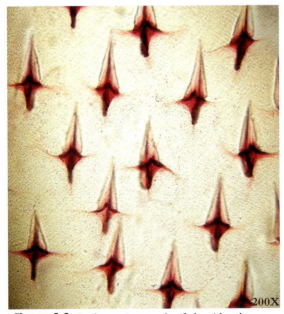

Figure 8.2 A photomicrograph of placoid scales.

200X

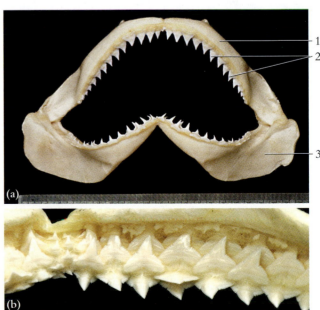

Figure 8.3 Shark jaws (a) and (b) a detailed view showing rows of replacement teeth (scale in mm).

1. Palatopterygoquadrate cartilage (upper jaw)
2. Placoid teeth
3. Meckel's cartilage (lower jaw)

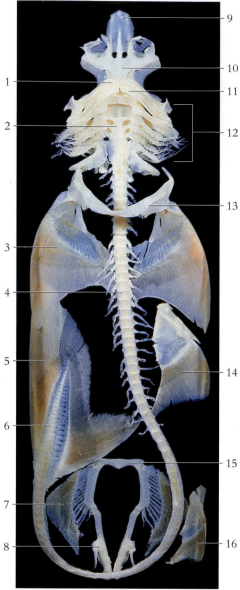

Figure 8.4 A ventral view of the cartilaginous skeleton of a male dogfish shark.

1. Palatopterygoquadrate cartilage (upper jaw)
2. Hypobranchial cartilage
3. Pectoral fin
4. Trunk vertebrae
5. Caudal fin
6. Caudal vertebrae
7. Pelvic fin
8. Clasper
9. Rostrum
10. Chondrocranium
11. Meckel's cartilage (lower jaw)
12. Visceral arches
13. Pectoral girdle
14. Anterior dorsal fin
15. Pelvic girdle
16. Posterior dorsal fin

Figure 8.5 The musculature of the jaw, gills, and pectoral fin of a dogfish shark.

1. 2nd dorsal constrictor
2. Levator of pectoral fin
3. 3rd through 6th ventral constrictors
4. Spiracular muscle
5. Facial nerve (hyomandibular branch)
6. Mandibular adductor
7. 2nd ventral constrictor

Figure 8.6 A ventral view of the hypobranchial musculature of the dogfish shark.

1. Depressor of pectoral fin
2. Common coracoarcual
3. Linea alba
4. Hypaxial muscle
5. 3rd through 6th ventral constrictors
6. 1st ventral constrictor
7. 2nd ventral constrictor
8. Mandibular adductor

Figure 8.7 A lateral view of the axial musculature of the dogfish shark.

1. Horizontal septum
2. Hypaxial myotome portion
3. Epaxial myotome portion
4. Lateral bundle of myotomes
5. Ventral bundle of myotomes

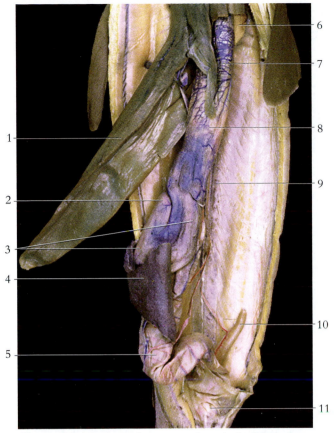

Figure 8.8 The internal anatomy of a male dogfish shark.

1. Right lobe of liver (reflected)
2. Pyloric sphincter valve
3. Stomach (pyloric region)
4. Spleen
5. Ileum
6. Testis
7. Esophagus
8. Stomach (cardiac region)
9. Kidney
10. Rectal gland
11. Cloaca

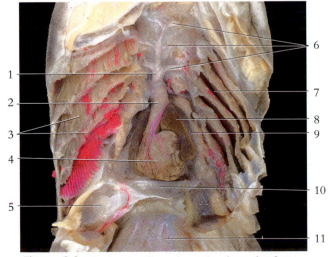

Figure 8.9 The heart, gills, and associated vessels of a dogfish shark.

1. Ventral aorta
2. Conus arteriosus
3. Gills
4. Ventricle
5. Pectoral girdle (cut)
6. Afferent branchial arteries
7. Gill cleft
8. Atrium
9. Pericardial cavity
10. Transverse septum
11. Liver

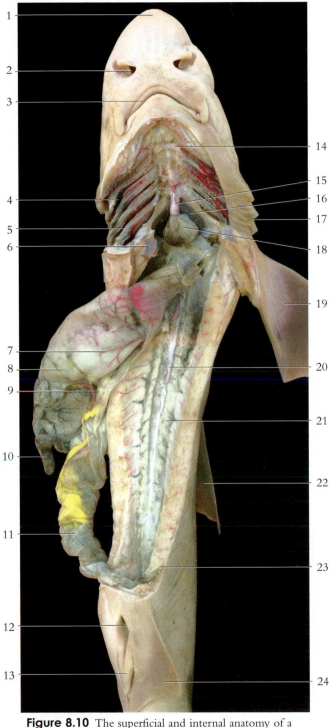

Figure 8.10 The superficial and internal anatomy of a dogfish shark (liver removed).

1. Rostrum
2. Nostril
3. Mouth
4. Gill cleft
5. Gill
6. Pectoral girdle (cut)
7. Gastroplenic artery
8. Stomach (reflected)
9. Pancreas
10. Spleen (reflected)
11. Intestine (reflected)
12. Cloaca
13. Clasper
14. Afferent branchial artery
15. Ventral aorta
16. Atrium
17. Gill slit
18. Heart
19. Pectoral fin
20. Dorsal aorta
21. Kidney
22. Dorsal fin
23. Rectal gland
24. Pelvic fin

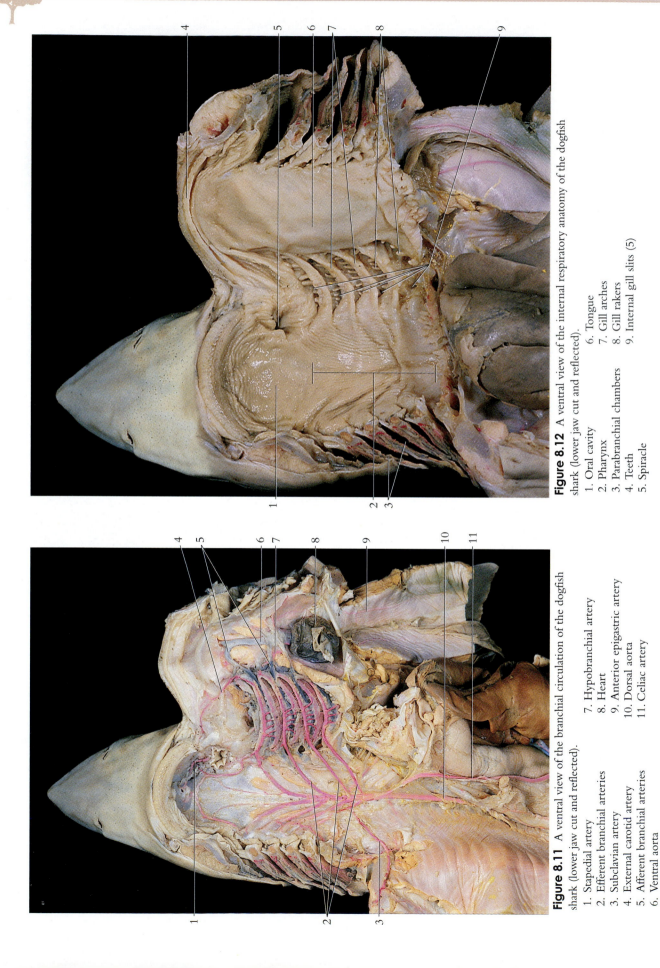

Figure 8.12 A ventral view of the internal respiratory anatomy of the dogfish shark (lower jaw cut and reflected).

1. Oral cavity
2. Pharynx
3. Parabranchial chambers
4. Teeth
5. Spiracle
6. Tongue
7. Gill arches
8. Gill rakers
9. Internal gill slits (5)

Figure 8.11 A ventral view of the branchial circulation of the dogfish shark (lower jaw cut and reflected).

1. Stapedial artery
2. Efferent branchial arteries
3. Subclavian artery
4. External carotid artery
5. Afferent branchial arteries
6. Ventral aorta
7. Hypobranchial artery
8. Heart
9. Anterior epigastric artery
10. Dorsal aorta
11. Celiac artery

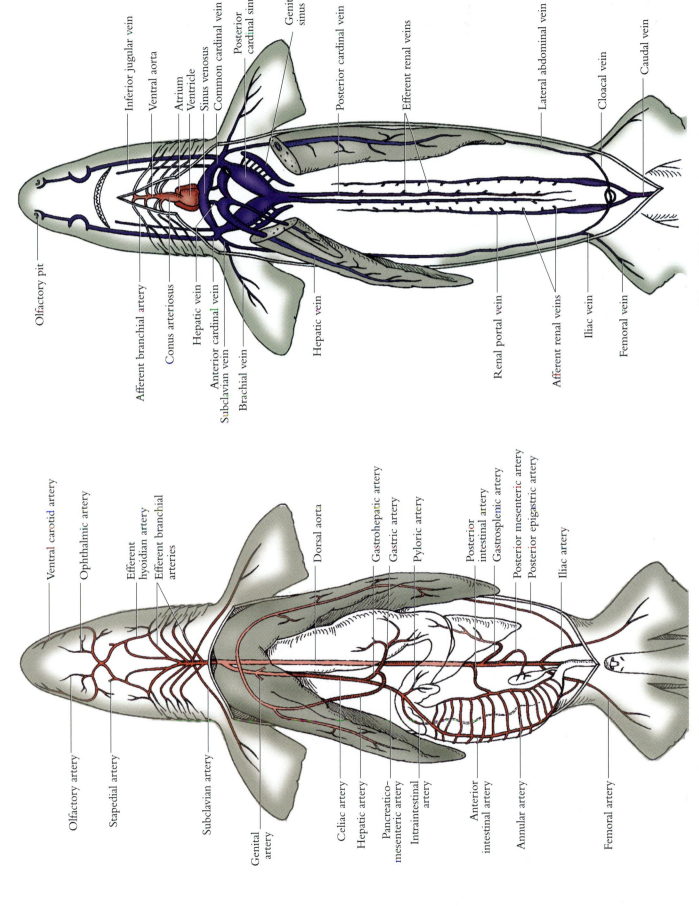

Figure 8.14 The veins of a dogfish shark.

Olfactory pit

Inferior jugular vein
Ventral aorta
Atrium
Ventricle
Sinus venosus
Common cardinal vein
Posterior cardinal sinus
Genital sinus

Posterior cardinal vein

Efferent renal veins

Lateral abdominal vein

Cloacal vein

Caudal vein

Afferent branchial artery
Conus arteriosus
Hepatic vein
Anterior cardinal vein
Subclavian vein
Brachial vein

Hepatic vein

Renal portal vein

Afferent renal veins

Iliac vein

Femoral vein

Figure 8.13 The arteries of a dogfish shark.

Ventral carotid artery
Ophthalmic artery
Efferent hyoidian artery
Efferent branchial arteries

Dorsal aorta

Gastrohepatic artery
Gastric artery
Pyloric artery

Posterior intestinal artery
Gastrosplenic artery
Posterior mesenteric artery
Posterior epigastric artery

Iliac artery

Olfactory artery

Stapedial artery

Subclavian artery

Genital artery

Celiac artery
Hepatic artery
Pancreatico-mesenteric artery
Intraintestinal artery

Anterior intestinal artery

Annular artery

Femoral artery

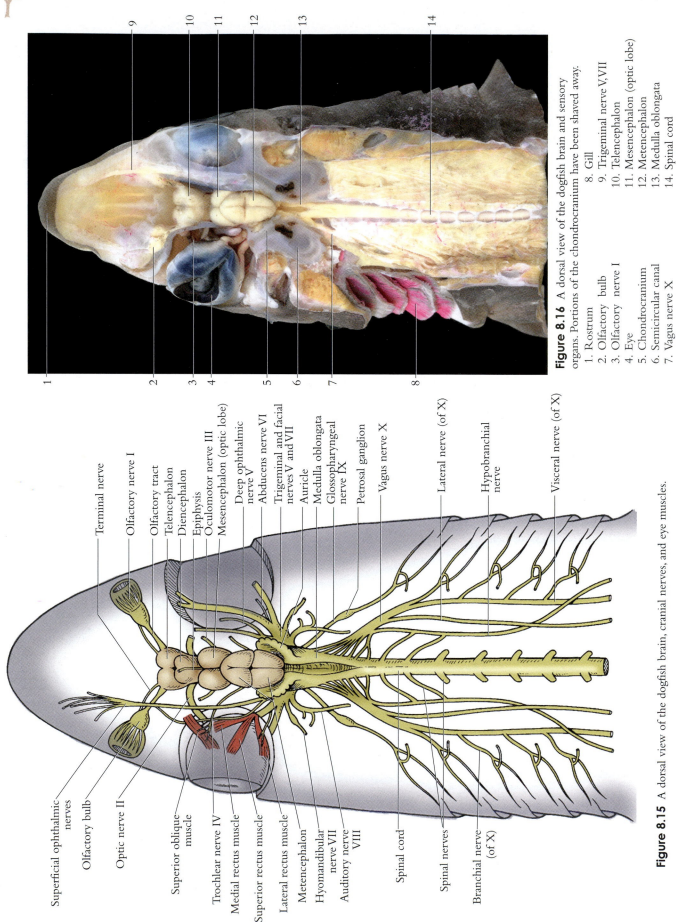

Figure 8.16 A dorsal view of the dogfish brain and sensory organs. Portions of the chondrocranium have been shaved away.

1. Rostrum
2. Olfactory bulb
3. Olfactory nerve I
4. Eye
5. Chondrocranium
6. Semicircular canal
7. Vagus nerve X
8. Gill
9. Trigeminal nerve V,VII
10. Telencephalon
11. Mesencephalon (optic lobe)
12. Metencephalon
13. Medulla oblongata
14. Spinal cord

Superficial ophthalmic nerves
Olfactory bulb
Optic nerve II
Superior oblique muscle
Trochlear nerve IV
Medial rectus muscle
Superior rectus muscle
Lateral rectus muscle
Metencephalon
Hyomandibular nerve VII
Auditory nerve VIII
Spinal cord
Spinal nerves
Branchial nerve (of X)

Terminal nerve
Olfactory nerve I
Olfactory tract
Telencephalon
Diencephalon
Epiphysis
Oculomotor nerve III
Mesencephalon (optic lobe)
Deep ophthalmic nerve V
Abducens nerve VI
Trigeminal and facial nerves V and VII
Auricle
Medulla oblongata
Glossopharyngeal nerve IX
Petrosal ganglion
Vagus nerve X
Lateral nerve (of X)
Hypobranchial nerve
Visceral nerve (of X)

Figure 8.15 A dorsal view of the dogfish brain, cranial nerves, and eye muscles.

Superclass Osteichthyes - Class Actinopterygii

Figure 8.17 The external anatomy of a perch.

1. Anterior dorsal fin
2. Eye
3. Nostrils
4. Mandible
5. Dentary
6. Operculum
7. Pectoral fin
8. Pelvic fin
9. Lateral line
10. Posterior dorsal fin
11. Caudal fin
12. Anal fin
13. Anus

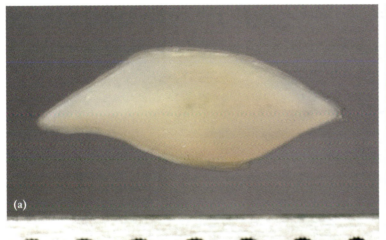

(a)

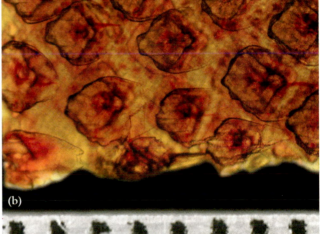

(b)

Figure 8.18 Ganoid scales, present in primitive fishes like the gar, are composed of silvery ganoin on the top surface and bone on the bottom. Two different sizes are shown (a) and (b) (scale in mm).

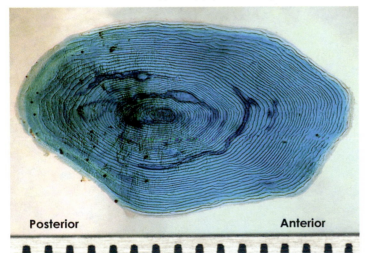

Posterior   Anterior

Figure 8.19 Cycloid scales, along with ctenoid scales (Figure 8.20), are found on advanced bony fishes. They are much thinner and more flexible than ganoid scales and overlap each other (scale in mm).

Posterior   Anterior

Figure 8.20 Ctenoid scales differ from cycloid scales in that they have comb-like ridges on the exposed edge, thought to improve swimming efficiency (scale in mm).

Figure 8.21 The skeleton of a perch.
1. Anterior dorsal fin
2. Fin spines
3. Neurocranium
4. Premaxilla
5. Maxilla
6. Dentary
7. Pelvic fin
8. Opercular bones
9. Pectoral fin
10. Vertebral column
11. Posterior dorsal fin
12. Soft rays
13. Caudal fin
14. Neural spine
15. Haemal spine
16. Anal fin
17. Ribs

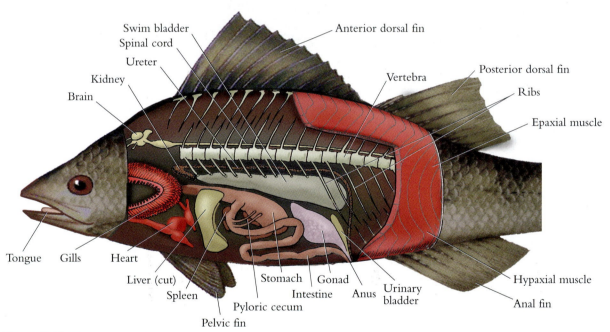

Figure 8.22 The anatomy of a perch.

Figure 8.23 The viscera of a perch.
1. Epaxial muscles
2. Stomach
3. Gill
4. Heart
5. Pyloric cecum
6. Liver (cut)
7. Pancreas
8. Vertebrae
9. Urinary bladder
10. Gonad
11. Anus
12. Intestine

Class Amphibia

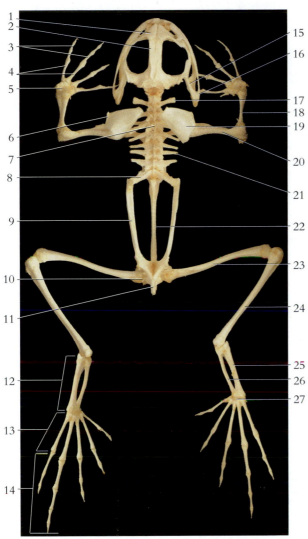

Figure 14.24 A dorsal view of the frog skeleton.

1. Nasal bone
2. Frontoparietal bone
3. Phalanges of digits
4. Metacarpal bones
5. Carpal bones
6. Scapula
7. Vertebra
8. Transverse process of sacral (ninth) vertebra
9. Ilium
10. Acetabulum
11. Ischium
12. Tarsal bones
13. Metatarsal bones

14. Phalanges of digits
15. Squamosal bone
16. Quadratojugal bone
17. Transverse process
18. Radioulna
19. Suprascapula
20. Humerus
21. Transverse process
22. Urostyle
23. Femur
24. Tibiofibula
25. Fibulare (calcaneum)
26. Tibiale (astragalus)
27. Distal tarsal bones

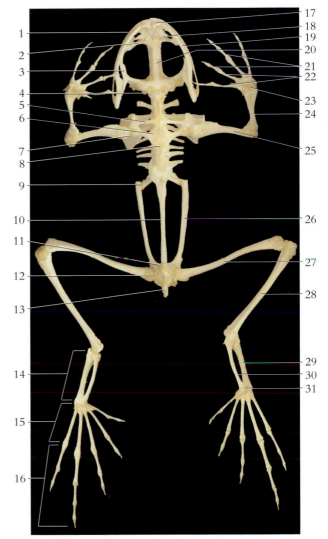

Figure 8.25 A ventral view of the frog skeleton.

1. Maxilla
2. Palatine
3. Pterygoid bone
4. Exoccipital bone
5. Clavicle
6. Coracoid
7. Glenoid fossa
8. Sternum
9. Transverse process of sacral (ninth) vertebra
10. Urostyle
11. Pubis
12. Acetabulum
13. Ischium
14. Tarsal bones
15. Metatarsal bones
16. Phalanges of digits

17. Premaxilla
18. Vomer
19. Dentary
20. Parasphenoid bone
21. Phalange of digits
22. Metacarpal bone
23. Carpal bones
24. Humerus
25. Radioulna
26. Ilium
27. Femur
28. Tibiofibula
29. Fibulare (calcaneum)
30. Tibiale (astragalus)
31. Distal tarsal bones

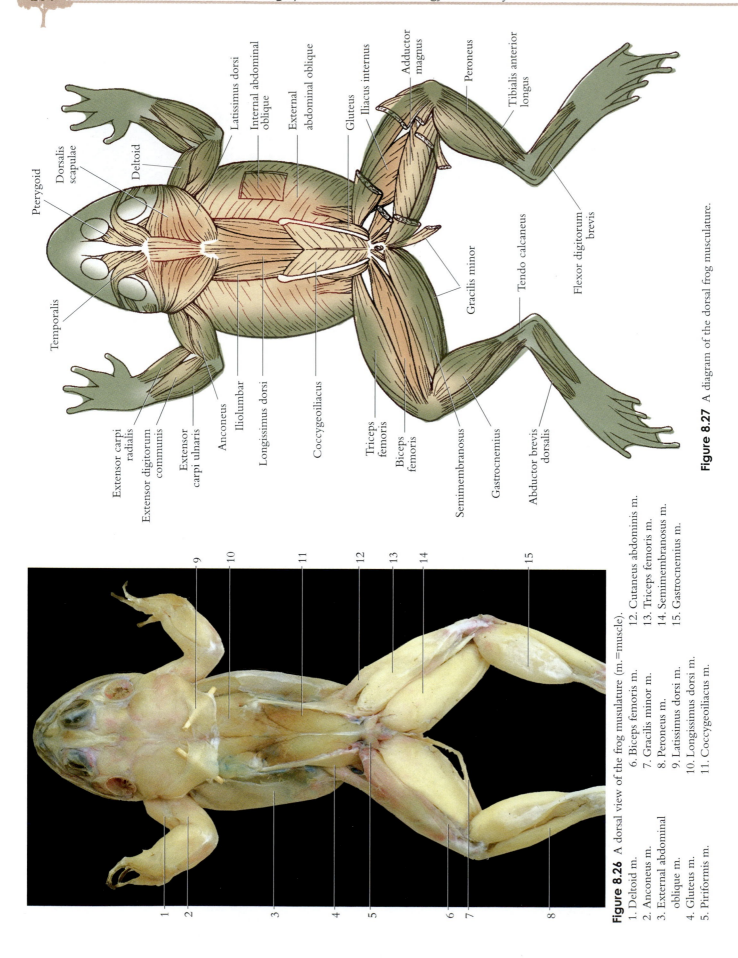

Figure 8.27 A diagram of the dorsal frog musculature.

Figure 8.26 A dorsal view of the frog musulature (m.=muscle).

1. Deltoid m.
2. Anconeus m.
3. External abdominal oblique m.
4. Gluteus m.
5. Piriformis m.

6. Biceps femoris m.
7. Gracilis minor m.
8. Peroneus m.
9. Latissimus dorsi m.
10. Longissimus dorsi m.
11. Coccygeoiliacus m.

12. Cutaneus abdominis m.
13. Triceps femoris m.
14. Semimembranosus m.
15. Gastrocnemius m.

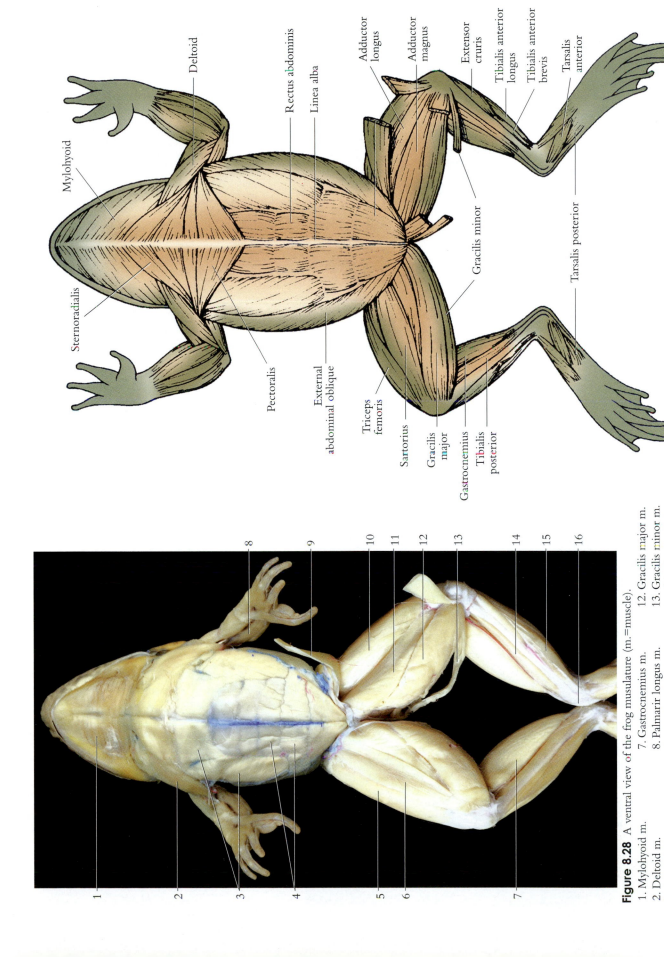

Figure 8.29 A diagram of the ventral frog musculature.

Mylohyoid

Deltoid

Rectus abdominis

Linea alba

Adductor longus

Adductor magnus

Extensor cruris

Tibialis anterior longus

Tibialis anterior brevis

Tarsalis anterior

Sternoradialis

Pectoralis

External abdominal oblique

Gracilis minor

Tarsalis posterior

Triceps femoris

Sartorius

Gracilis major

Gastrocnemius

Tibialis posterior

Figure 8.28 A ventral view of the frog musculature (m.=muscle).

1. Mylohyoid m.
2. Deltoid m.
3. Pectoralis m.
4. Rectus abdominis m.
5. Triceps femoris m.
6. Sartorius m.

7. Gastrocnemius m.
8. Palmarir longus m.
9 Sartorius m. (cut and reflected)
10. Adductor longus m.
11. Adductor magnus m.

12. Gracilis major m.
13. Gracilis minor m.
14. Tibialis posterior m.
15. Tibialis anterior m.
16. Tendo calcaneus

Figure 8.30 A dorsal view of the leg muscles of a frog (m.=muscle).

1. Gluteus m.
2. Cutaneus abdominis m.
3. Piriformis m.
4. Semimembranosus m. (cut and reflected)
5. Gracilis minor m.
6. Peroneus m.
7. Coccygeoiliacus m.
8. Triceps femoris m. (cut and reflected)
9. Iliacus internus m.
10. Biceps femoris m.
11. Adductor magnus m.
12. Semitendinosus m . (cut and reflected)
13. Gastrocnemius m.

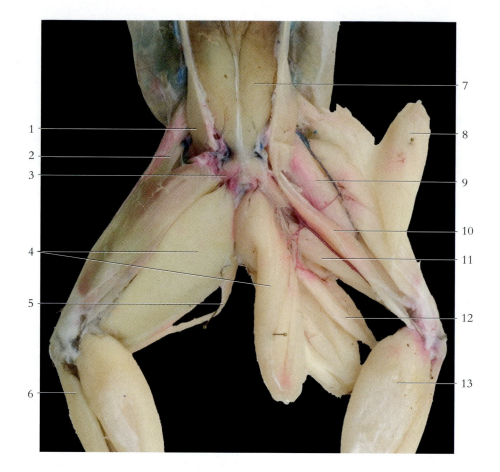

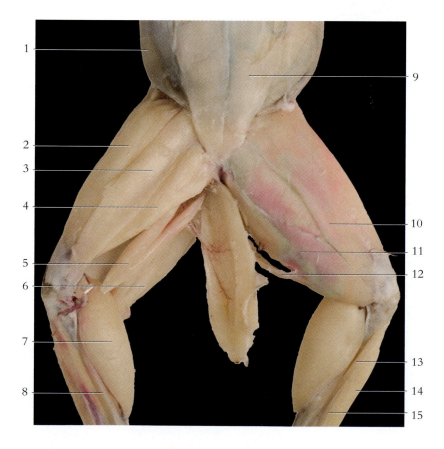

Figure 8.31 A ventral view of the leg muscles of a frog (m.=muscle).

1. External abdominal oblique m.
2. Triceps femoris m.
3. Adductor longus m.
4. Adductor magnus m.
5. Semitendinosus m. (cut)
6. Semimembranosus m.
7. Gastrocnemius m.
8. Tibialis posterior m.
9. Rectus abdominis m.
10. Sartorius m.
11. Gracilis major m.
12. Gracilis minor m.
13. Extensor cruris m.
14. Tibialis anterior longus m.
15. Tibialis anterior brevis m.

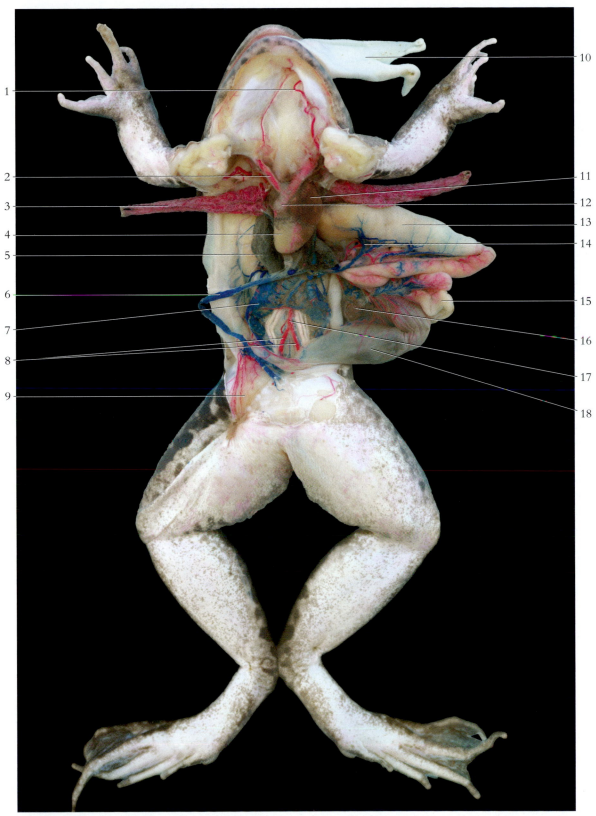

Figure 8.32 The internal anatomy of the frog.

1. External carotid artery
2. Truncus arteriosus
3. Lung (reflected)
4. Ventricle of heart
5. Liver (cut)
6. Ventral abdominal vein
7. Kidney
8. Sciatic arteries
9. Bladder
10. Tongue
11. Atrium of heart
12. Conus arteriosus
13. Stomach (reflected)
14. Gastric vein
15. Small intestine
16. Spleen
17. Dorsal aorta
18. Large intestine

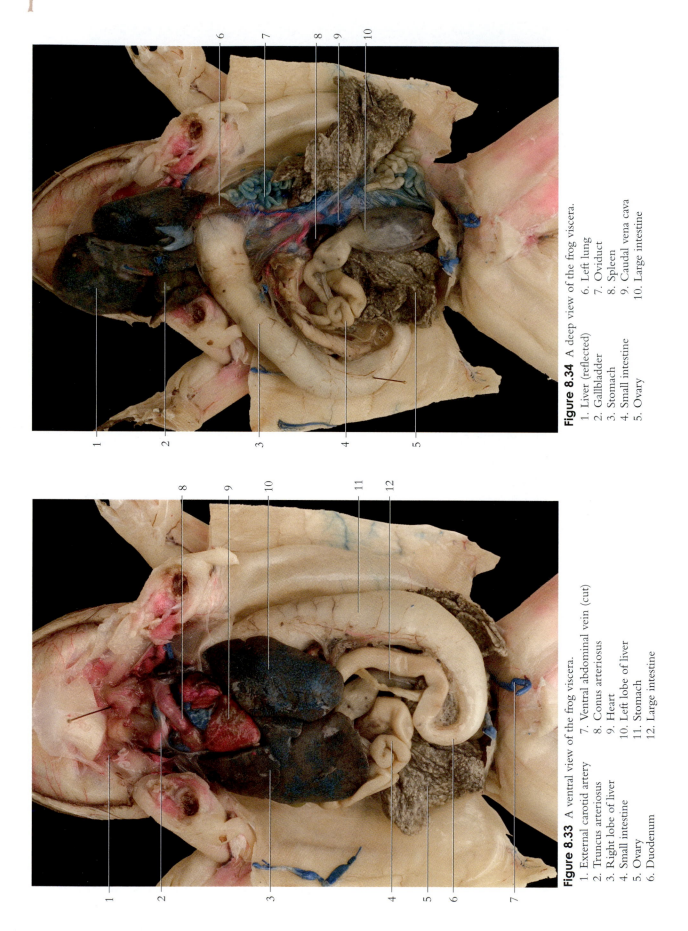

Figure 8.33 A ventral view of the frog viscera.

1. External carotid artery
2. Truncus arteriosus
3. Right lobe of liver
4. Small intestine
5. Ovary
6. Duodenum
7. Ventral abdominal vein (cut)
8. Conus arteriosus
9. Heart
10. Left lobe of liver
11. Stomach
12. Large intestine

Figure 8.34 A deep view of the frog viscera.

1. Liver (reflected)
2. Gallbladder
3. Stomach
4. Small intestine
5. Ovary
6. Left lung
7. Oviduct
8. Spleen
9. Caudal vena cava
10. Large intestine

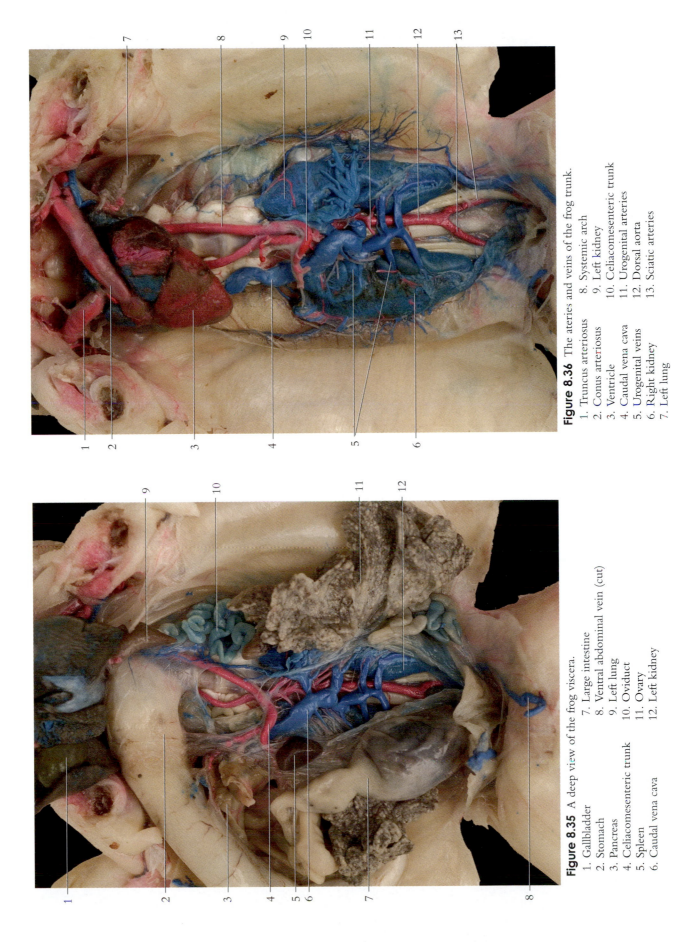

Figure 8.36 The ateries and veins of the frog trunk.

1. Truncus arteriosus
2. Conus arteriosus
3. Ventricle
4. Caudal vena cava
5. Urogenital veins
6. Right kidney
7. Left lung
8. Systemic arch
9. Left kidney
10. Celiacomesenteric trunk
11. Urogenital arteries
12. Dorsal aorta
13. Sciatic arteries

Figure 8.35 A deep view of the frog viscera.

1. Gallbladder
2. Stomach
3. Pancreas
4. Celiacomesenteric trunk
5. Spleen
6. Caudal vena cava
7. Large intestine
8. Ventral abdominal vein (cut)
9. Left lung
10. Oviduct
11. Ovary
12. Left kidney

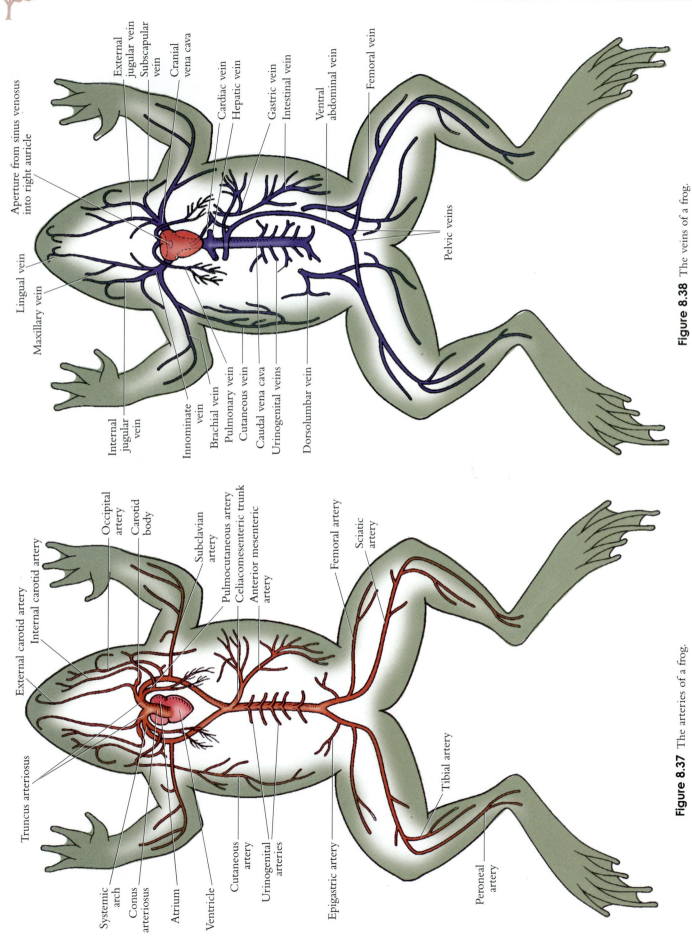

Figure 8.38 The veins of a frog.

Figure 8.37 The arteries of a frog.

Class Sauropsida (=Reptilia)

Figure 8.39 A dorsal view of a turtle.

1. Eye
2. Pentadactyl foot
3. Vertebral scales
4. Marginal scales
   (encircle the carapace)
5. Nostril
6. Head
7. Nuchal scale
8. Costal scales

Figure 8.40 A ventral view of a turtle.

1. Gular scales
2. Humeral scales
3. Pectoral scales
4. Abdominal scales
5. Femoral scales
6. Anal scales
7. Tail

Figure 8.41 The skull of a turtle.

1. Parietal bone
2. Supraoccipital bone
3. Postorbital bone
4. Jugal bone
5. Quadratojugal bone
6. Exoccipital bone
7. Quadrate bone
8. Supraangular bone
9. Articular bone
10. Angular bone
11. Frontal bone
12. Prefrontal bone
13. Palatine bone
14. Premaxilla
15. Maxilla
16. Beak
17. Dentary

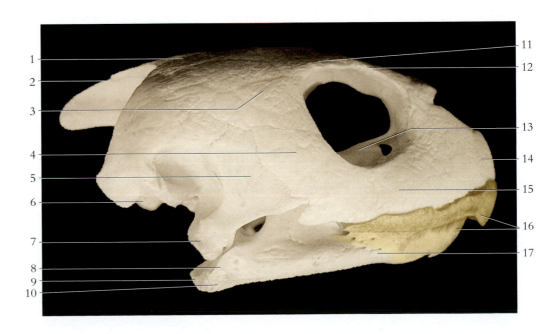

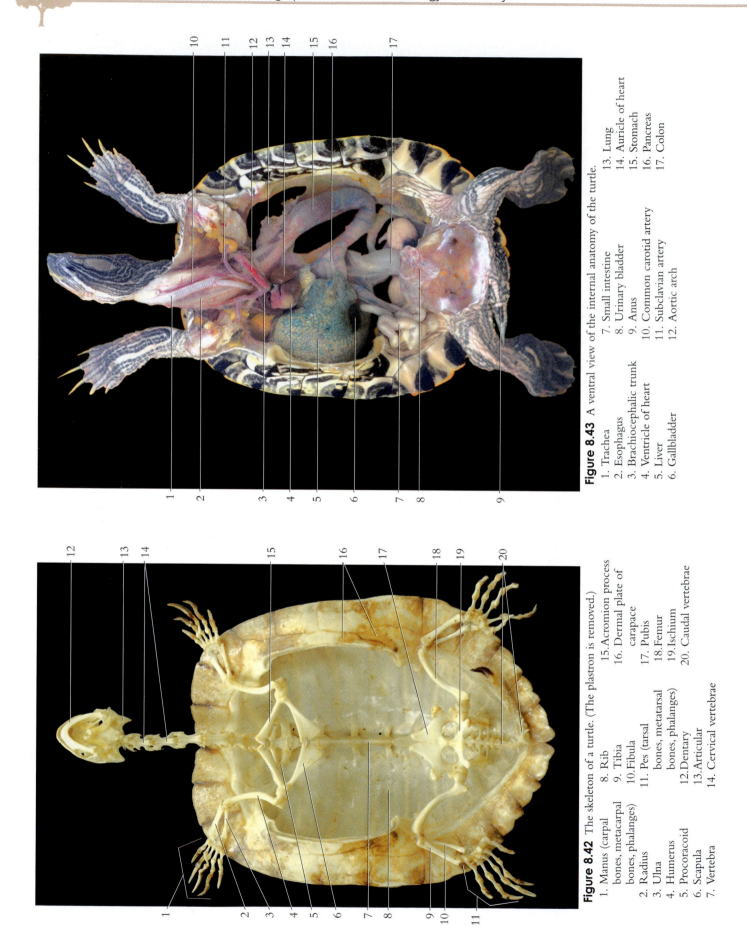

Figure 8.43 A ventral view of the internal anatomy of the turtle.

1. Trachea	7. Small intestine	13. Lung
2. Esophagus	8. Urinary bladder	14. Auricle of heart
3. Brachiocephalic trunk	9. Anus	15. Stomach
4. Ventricle of heart	10. Common carotid artery	16. Pancreas
5. Liver	11. Subclavian artery	17. Colon
6. Gallbladder	12. Aortic arch	

Figure 8.42 The skeleton of a turtle. (The plastron is removed.)

1. Manus (carpal bones, metacarpal bones, phalanges)	8. Rib	15. Acromion process
2. Radius	9. Tibia	16. Dermal plate of carapace
3. Ulna	10. Fibula	17. Pubis
4. Humerus	11. Pes (tarsal bones, metatarsal bones, phalanges)	18. Femur
5. Procoracoid	12. Dentary	19. Ischium
6. Scapula	13. Articular	20. Caudal vertebrae
7. Vertebra	14. Cervical vertebrae	

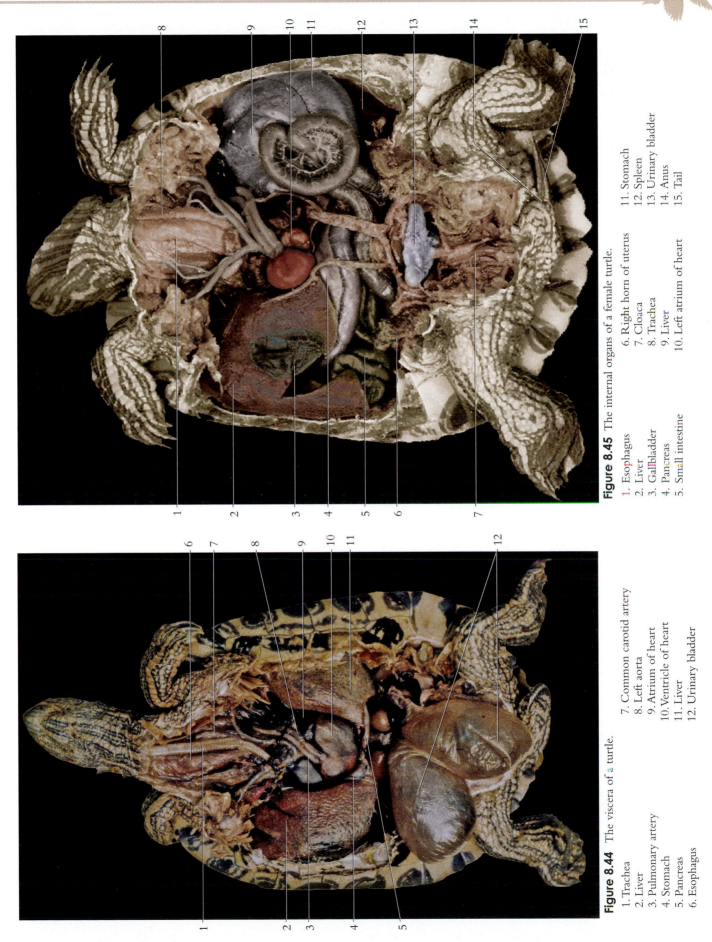

Figure 8.45 The internal organs of a female turtle.

1. Esophagus
2. Liver
3. Gallbladder
4. Pancreas
5. Small intestine
6. Right horn of uterus
7. Cloaca
8. Trachea
9. Liver
10. Left atrium of heart
11. Stomach
12. Spleen
13. Urinary bladder
14. Anus
15. Tail

Figure 8.44 The viscera of a turtle.

1. Trachea
2. Liver
3. Pulmonary artery
4. Stomach
5. Pancreas
6. Esophagus
7. Common carotid artery
8. Left aorta
9. Atrium of heart
10. Ventricle of heart
11. Liver
12. Urinary bladder

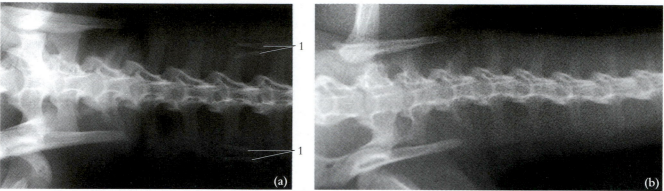

Figure 8.46 Male lizards and snakes have hemipenes as copulatory organs. The hemipene seen in a radiograph of a male (a) crocodile monitor, *Varanus salvadorii*. As seen in a radiograph, a female (b) lacks a hemipene. The female cloaca is the receptacle of the everted male hemipene during copulation.
 1. Sheaths of hemipenes

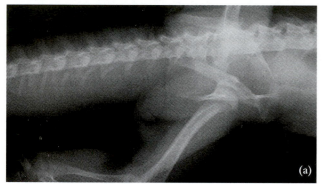
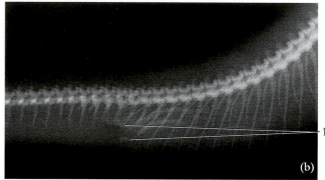

Figure 8.47 A radiograph of the pelvic region of a savannah monitor (a) showing a highly developed limb. Compare this to the radiograph of the pelvic region of a boa (b) showing the vestigeal pelvic girdle.
 1. Vestigeal pelvic girdle

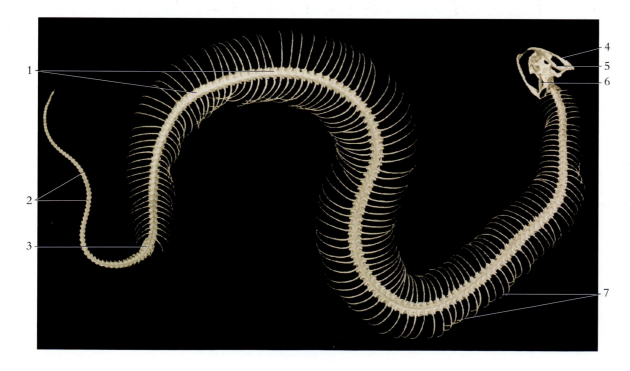

Figure 8.48 The skeleton of a snake (python).
 1. Caudal vertebrae
 2. Vestigial pelvic girdle
 3. Trunk vertebrae
 4. Ribs
 5. Dentary
 6. Quadrate bone
 7. Supratemporal bone

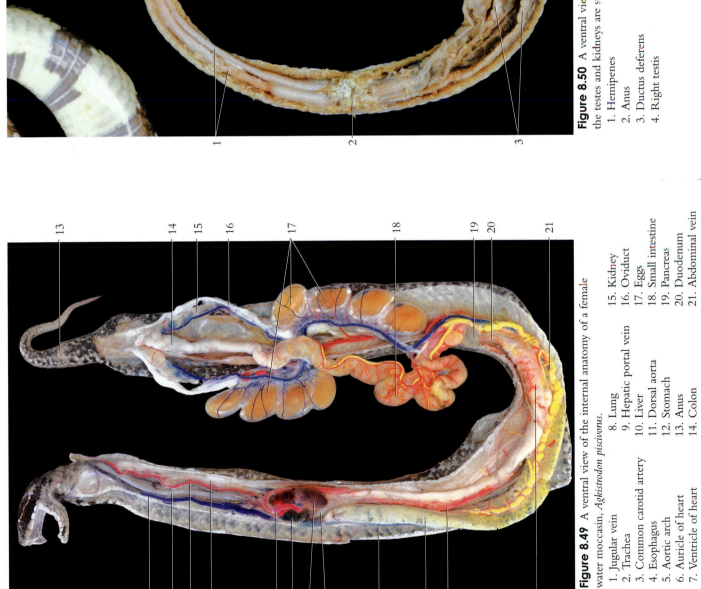

Figure 8.50 A ventral view of the internal anatomy of a male snake. Note that the testes and kidneys are staggered.

1. Hemipenes
2. Anus
3. Ductus deferens
4. Right testis
5. Intestine
6. Left testis
7. Ductus deferens
8. Right kidney
9. Ureter
10. Left kidney

Figure 8.49 A ventral view of the internal anatomy of a female water moccasin, *Agkistrodon piscivorus*.

1. Jugular vein
2. Trachea
3. Common carotid artery
4. Esophagus
5. Aortic arch
6. Auricle of heart
7. Ventricle of heart
8. Lung
9. Hepatic portal vein
10. Liver
11. Dorsal aorta
12. Stomach
13. Anus
14. Colon
15. Kidney
16. Oviduct
17. Eggs
18. Small intestine
19. Pancreas
20. Duodenum
21. Abdominal vein

Class Aves

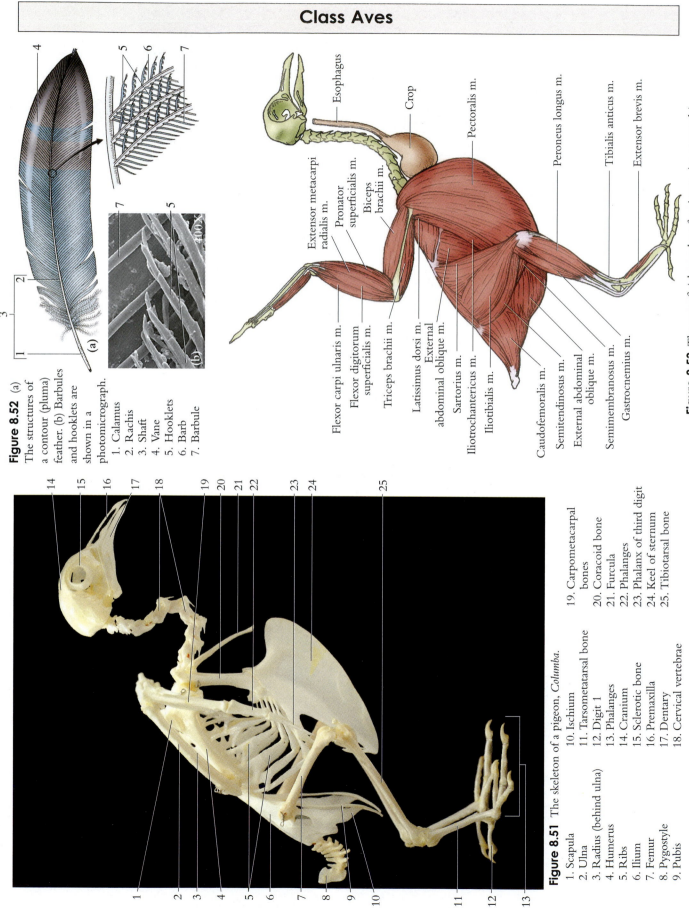

Figure 8.52 (a) The structures of a contour (pluma) feather. (b) Barbules and hooklets are shown in a photomicrograph.

1. Calamus
2. Rachis
3. Shaft
4. Vane
5. Hooklets
6. Barb
7. Barbule

Figure 8.53 The superficial muscles of a pigeon. (m. = muscle)

Figure 8.51 The skeleton of a pigeon, *Columba*.

1. Scapula
2. Ulna
3. Radius (behind ulna)
4. Humerus
5. Ribs
6. Ilium
7. Femur
8. Pygostyle
9. Pubis
10. Ischium
11. Tarsometatarsal bone
12. Digit 1
13. Phalanges
14. Cranium
15. Sclerotic bone
16. Premaxilla
17. Dentary
18. Cervical vertebrae
19. Carpometacarpal bones
20. Coracoid bone
21. Furcula
22. Phalanges
23. Phalanx of third digit
24. Keel of sternum
25. Tibiotarsal bone

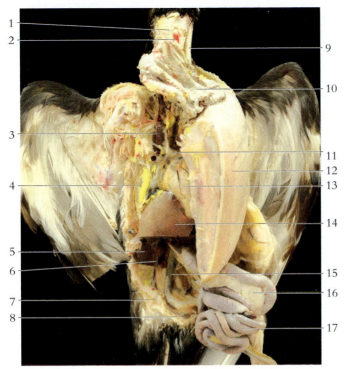

Figure 8.54 A ventral view of the internal anatomy of a pigeon.

1. Esophagus
2. Carotid artery
3. Vena cava
4. Lung
5. Kidney
6. Oviduct
7. Oil gland
8. Cloaca
9. Trachea
10. Crop
11. Aortic arch
12. Pectoralis muscle
13. Heart
14. Liver (cut)
15. Rectum
16. Pancreas
17. Ileum

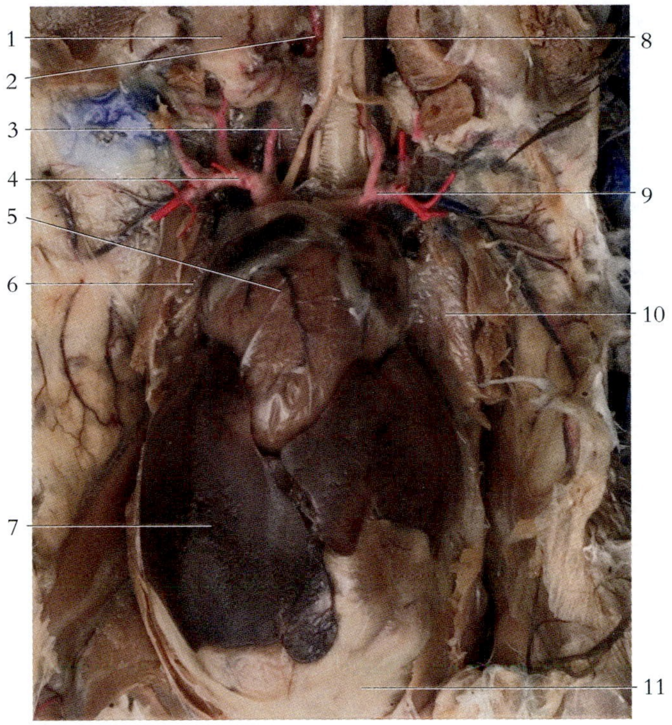

Figure 8.55 A ventral view of a pigeon heart and surrounding organs.

1. Crop
2. Common carotid artery
3. Esophagus
4. Right subclavian artery
5. Heart (within pericardium)
6. Right lung
7. Liver
8. Trachea
9. Left subclavian artery
10. Left lung
11. Greater omentum

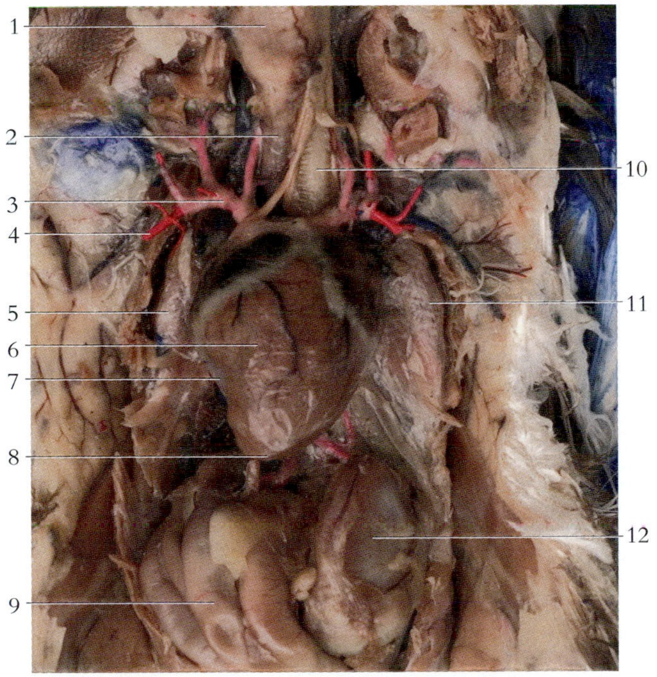

Figure 8.56 A ventral view of the viscera of a pigeon, *Columba,* with the liver removed.

1. Crop
2. Esophagus
3. Right subclavian artery
4. Axillary artery
5. Right lung
6. Heart
7. Pericardium
8. Apex of heart
9. Small intestine
10. Trachea
11. Left lung
12. Gizzard

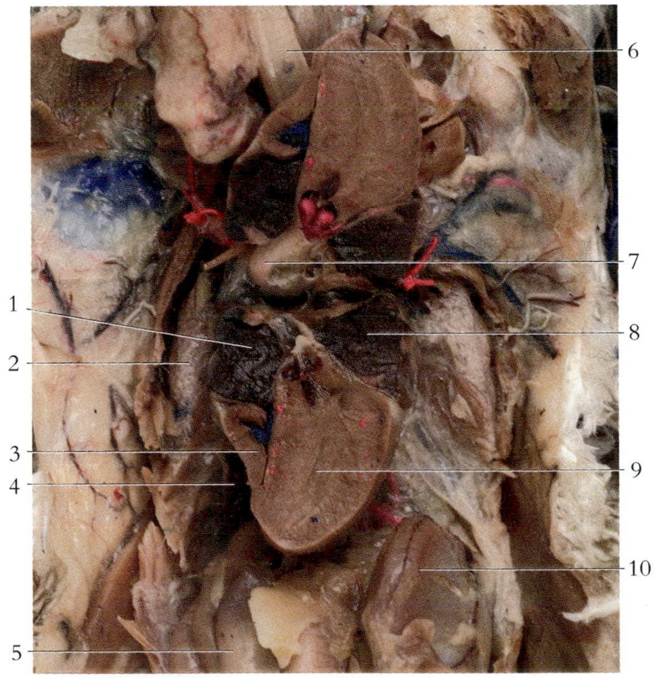

Figure 8.57 The viscera of a pigeon with the heart sectioned.

1. Right atrium
2. Right lung
3. Right ventricle
4. Liver (cut)
5. Small intestine
6. Trachea
7. Aortic arch
8. Left atrium
9. Left ventricle
10. Gizzard

Class Mammalia

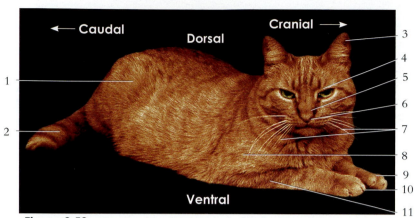

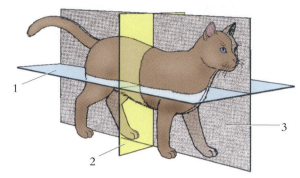

Figure 8.58 Directional terminology and superficial structures in a cat (quadrupedal vertebrate).

1. Thigh
2. Tail
3. Auricle (pinna)
4. Superior palpebra
   (superior eyelid)
5. Bridge of nose
6. Naris (nostril)
7. Vibrissae
8. Brachium
9. Manus (front foot)
10. Claw
11. Antebrachium

Figure 8.59 The planes of reference in a cat.
1. Coronal plane (frontal plane)
2. Transverse plane (cross-sectional plane)
3. Midsagittal plane (median plane)

Rat Dissection

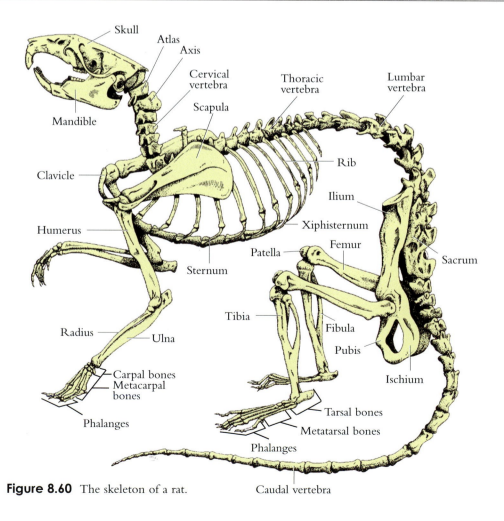

Figure 8.60 The skeleton of a rat.

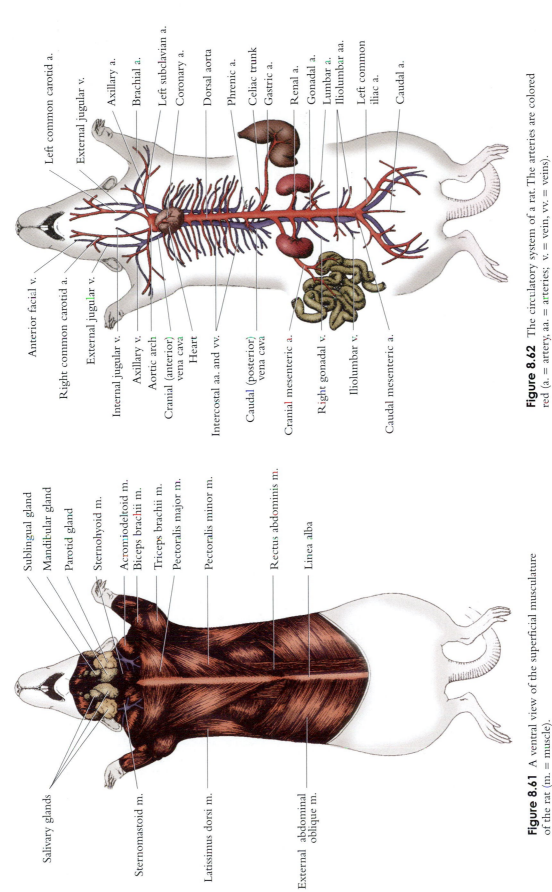

Left common carotid a.
External jugular v.
Axillary a.
Brachial a.
Left subclavian a.
Coronary a.
Dorsal aorta
Phrenic a.
Celiac trunk
Gastric a.
Renal a.
Gonadal a.
Lumbar a.
Iliolumbar aa.
Left common iliac a.
Caudal a.

Anterior facial v.
Right common carotid a.
External jugular v.
Internal jugular v.
Axillary v.
Aortic arch
Cranial (anterior) vena cava
Heart
Intercostal aa. and vv.
Caudal (posterior) vena cava
Cranial mesenteric a.
Right gonadal v.
Iliolumbar v.
Caudal mesenteric a.

Figure 8.62 The circulatory system of a rat. The arteries are colored red (a. = artery, aa. = arteries; v. = vein, vv. = veins).

Sublingual gland
Mandibular gland
Parotid gland
Sternohyoid m.
Acromiodeltoid m.
Biceps brachii m.
Triceps brachii m.
Pectoralis major m.
Pectoralis minor m.
Rectus abdominis m.
Linea alba

Salivary glands
Sternomastoid m.
Latissimus dorsi m.
External abdominal oblique m.

Figure 8.61 A ventral view of the superficial musculature of the rat (m. = muscle).

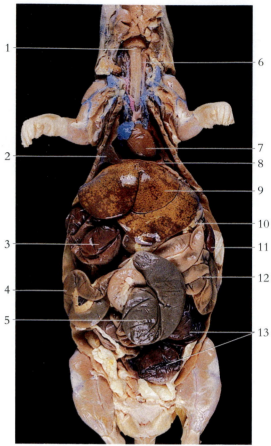

Figure 8.63 A ventral view of the rat viscera.

1. Trachea
2. Right lung
3. Right uterine horn of pregnant female
4. Jejunum
5. Cecum
6. Esophagus
7. Heart
8. Diaphragm (cut)
9. Liver
10. Stomach
11. Spleen
12. Ileum
13. Left uterine horn of pregnant female

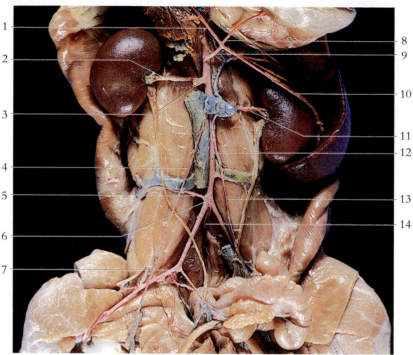

Figure 8.64 The abdominal arteries of the rat.

1. Hepatic a.
2. Right renal a.
3. Cranial mesentric a.
4. Right testicular a.
5. Right iliolumbar a.
6. Caudal mesenteric a. (cut)
7. Right common iliac a.
8. Gastric a.
9. Celiac trunk
10. Splenic a.
11. Left renal a.
12. Abdominal aorta
13. Left testicular a.
14. Middle sacral a.

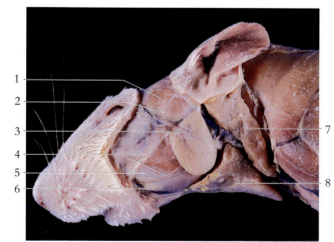

Figure 8.65 The head and neck region of the rat.

1. Temporalis muscle
2. Extraorbital lacrimal duct
3. Extraorbital lacrimal gland
4. Facial nerve
5. Masseter muscle
6. Parotid duct
7. Parotid gland
8. Mandibular gland

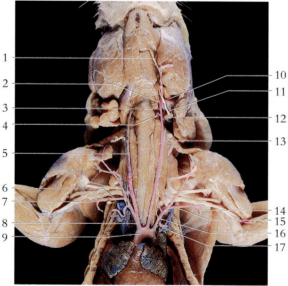

Figure 8.66 The arteries of the thoracic and neck regions of the rat.

1. Facial a.
2. Lingual a.
3. External carotid a.
4. Cranial thyroid a.
5. Common carotid a.
6. Axillary a.
7. Brachial a.
8. Brachiocephalic a.
9. Aortic arch
10. External maxillary a.
11. Internal carotid a.
12. Occipital a.
13. Common carotid a.
14. Lateral thoracic a.
15. Axillary a.
16. Subclavian a.
17. Internal thoracic a.

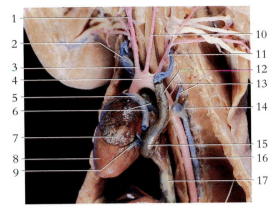

Figure 8.67 A reflected rat heart showing the major veins and arteries.

1. Right common carotid a.
2. Right cranial vena cava
3. Brachiocephalic trunk
4. Aortic arch
5. Pulmonary trunk
6. Left and right pulmonary aa.
7. Left auricle
8. Left ventricle
9. Coronary vein
10. Left common carotid a.
11. Left subclavian a.
12. Left cranial vena cava
13. Intercostal a. and v.
14. Azygos v.
15. Coronary sinus
16. Aorta
17. Caudal vena cava

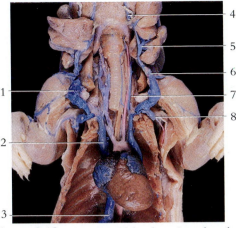

Figure 8.68 The veins of the thoracic and neck regions of the rat.

1. Cephalic v.
2. Cranial vena cava
3. Caudal vena cava
4. Linguofacial v.
5. Maxillary v.
6. External jugular v.
7. Internal jugular v.
8. Lateral thoracic v.

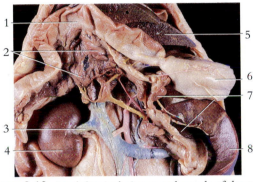

Figure 8.69 An abdominal viscera and vessels of the rat.

1. Duodenum
2. Biliary and duodenal parts of pancreas
3. Right renal v.
4. Right kidney
5. Liver (cut)
6. Stomach
7. Gastrosplenic part of pancreas
8. Spleen

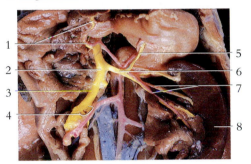

Figure 8.70 The branches of the hepatic portal system.

1. Cranial pancreaticoduodenal v.
2. Hepatic portal v.
3. Cranial mesenteric v.
4. Intestinal branches
5. Gastric v.
6. Gastrosplenic v.
7. Splenic branches
8. Spleen

Figure 8.71 The urogenital system of the male rat.

1. Vesicular gland
2. Prostate (dorsolateral part)
3. Prostate (ventral part)
4. Urethra in the pelviccanal
5. Ductus (vas) deferens
6. Crus of penis (cut)
7. Head of epididymis
8. Testis
9. Tail of epididymis
10. Urinary bladder
11. Symphysis pubis (cut exposing pelvic canal)
12. Bulbourethral glands
13. Bulbocavernosus muscle
14. Penis

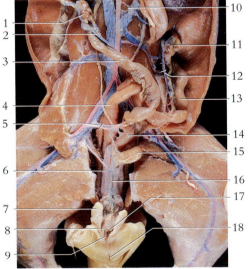

Figure 8.72 The urogenital system of the female rat.

1. Ovary
2. Uterine a. and v.
3. Uterine horn
4. Colon
5. Vesicular a. (umbilical a.)
6. Vagina
7. Preputial gland
8. Clitoris
9. Vaginal opening
10. Ovarian a. and v.
11. Ovary
12. Uterine a. and v.
13. Uterine horn
14. Uterine body
15. Urinary bladder
16. Urethra
17. Urethral opening
18. Anus

Fetal Pig Dissection

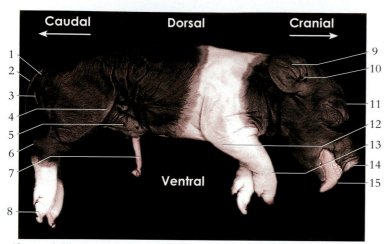

Figure 8.73 The directional terminology and superficial structures in a fetal pig (quadrupedal vertebrate).

1. Anus
2. Tail
3. Scrotum
4. Knee
5. Teat
6. Ankle
7. Umbilical cord
8. Hoof
9. Auricle (pinna)
10. External auditory canal
11. Superior palpebra(superior eyelid)
12. Elbow
13. Wrist
14. Naris (nostril)
15. Tongue

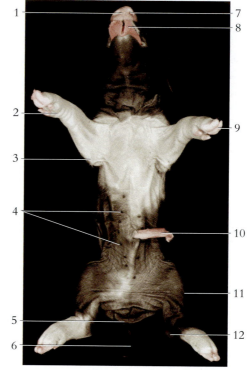

Figure 8.74
A ventral view of the surface anatomy of the fetal pig.
1. Nose
2. Wrist
3. Elbow
4. Teats
5. Scrotum
6. Tail
7. Nostril
8. Tongue
9. Digit
10. Umbilical cord
11. Knee
12. Ankle

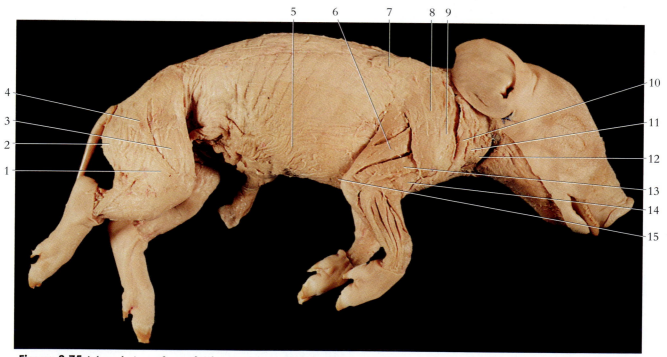

Figure 8.75 A lateral view of superficial musculature of the fetal pig.

1. Biceps femoris m.
2. Semitendinosus m.
3. Tensor fasciae latae m.
4. Gluteus medius m.
5. External abdominal oblique m.
6. Triceps brachii m. (long head)
7. Trapezius m.
8. Deltoid m.
9. Supraspinatus m.
10. Cleidooccipitalis m.
11. Cleidomastoid m.
12. Sternocephalicus m.
13. Triceps brachii m. (lateral head)
14. Brachialis m.
15. Pectoralis profundus m.

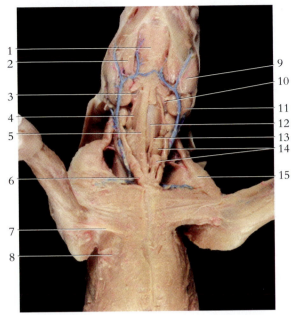

Figure 8.76 A ventral view of superficial muscles of neck and upper torso.

1. Mylohyoid m.
2. Digastric m.
3. Stylohyoid m.
4. Omohyoid m.
5. Sternohyoid m.
6. Sternomastoid m.
7. Pectoralis superficialis m.
8. Pectoralis profundus m.

9. Masseter m.
10. Thyrohyoid m.
11. Mandibular gland
12. Larynx
13. Sternothyroid m.
14. Mandibular lymph nodes
15. Brachiocephalic m.

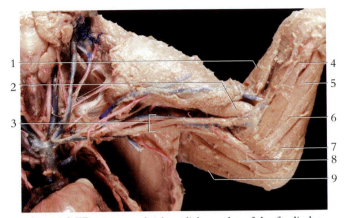

Figure 8.77 The superficial medial muscles of the forelimb.

1. Extensor carpi radialis m.
2. Biceps brachii m.
3. Brachial artery and vein, brachial plexus
4. Flexor carpi radialis m.
5. Flexor digitorum profundus m.

6. Flexor digitorum superficialis m.
7. Flexor carpi ulnaris m.
8. Triceps brachii m. (lateral head)
9. Triceps brachii m. (long head)

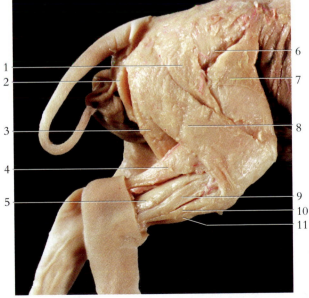

Figure 8.78 A lateral view of the superficial thigh and leg.

1. Gluteus superficialis m.
2. Semitendinosus m.
3. Semimembranosus m.
4. Gastrocnemius m.
5. Extensor digitorum quarti and quinti mm.
6. Gluteus medius m.

7. Tensor fasciae latae m.
8. Biceps femoris m.
9. Fibulais (peroneus) longus m.
10. Fibulais (peroneus) tertius m.
11. Tibialis anterior m.

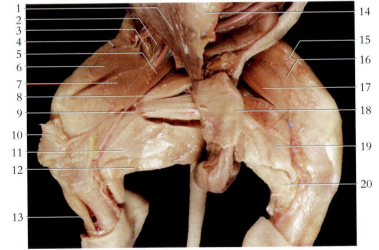

Figure 8.79 The medial muscles of thigh and leg.

1. External abdominal oblique m.
2. Psoas major m.
3. Iliacus m.
4. Tensor fasciae latae m.
5. Sartorius m.
6. Rectus femoris m.
7. Vastus medialis m.
8. Pectineus m.
9. Adductor m.
10. Aponeurosis of gracilis (cut)

11. Semimembranosus m.
12. Semitendinosus m.
13. Tibialis anterior m.
14. Linea alba
15. Rectus femoris m.
16. Vastus medialis m.
17. Sartorius m.
18. Gracilis m. (cut)
19. Gracilis m.
20. Semitendinosus m.

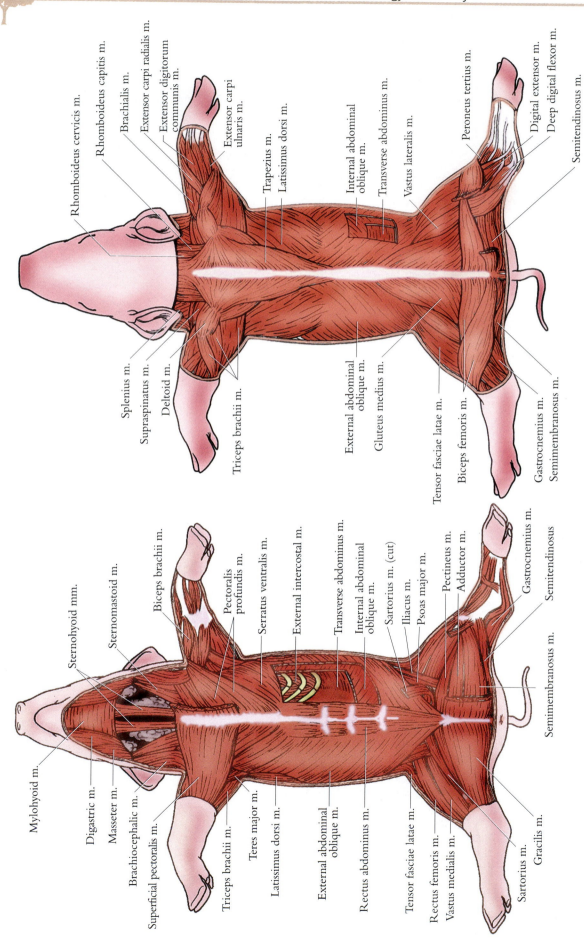

Rhomboideus cervicis m.
Rhomboideus capitis m.
Brachialis m.
Extensor carpi radialis m.
Extensor digitorum communis m.
Extensor carpi ulnaris m.
Trapezius m.
Latissimus dorsi m.
Internal abdominal oblique m.
Transverse abdominus m.
Vastus lateralis m.
Peroneus tertius m.
Digital extensor m.
Deep digital flexor m.
Semitendinosus m.

Splenius m.
Supraspinatus m.
Deltoid m.
Triceps brachii m.
External abdominal oblique m.
Gluteus medius m.
Tensor fasciae latae m.
Biceps femoris m.
Gastrocnemius m.
Semimembranosus m.

Figure 8.81   A dorsal view of the muscles of the fetal pig.

Sternohyoid mm.
Sternomastoid m.
Biceps brachii m.
Pectoralis profundis m.
Serratus ventralis m.
External intercostal m.
Transverse abdominus m.
Internal abdominal oblique m.
Sartorius m. (cut)
Iliacus m.
Psoas major m.
Pectineus m.
Adductor m.
Gastrocnemius m.
Semitendinosus

Mylohyoid m.
Digastric m.
Masseter m.
Brachiocephalic m.
Superficial pectoralis m.
Triceps brachii m.
Teres major m.
Latissimus dorsi m.
External abdominal oblique m.
Rectus abdominus m.
Tensor fasciae latae m.
Rectus femoris m.
Vastus medialis m.
Sartorius m.
Gracilis m.
Semimembranosus m.

Figure 8.80   A ventral view of the muscles of the fetal pig.

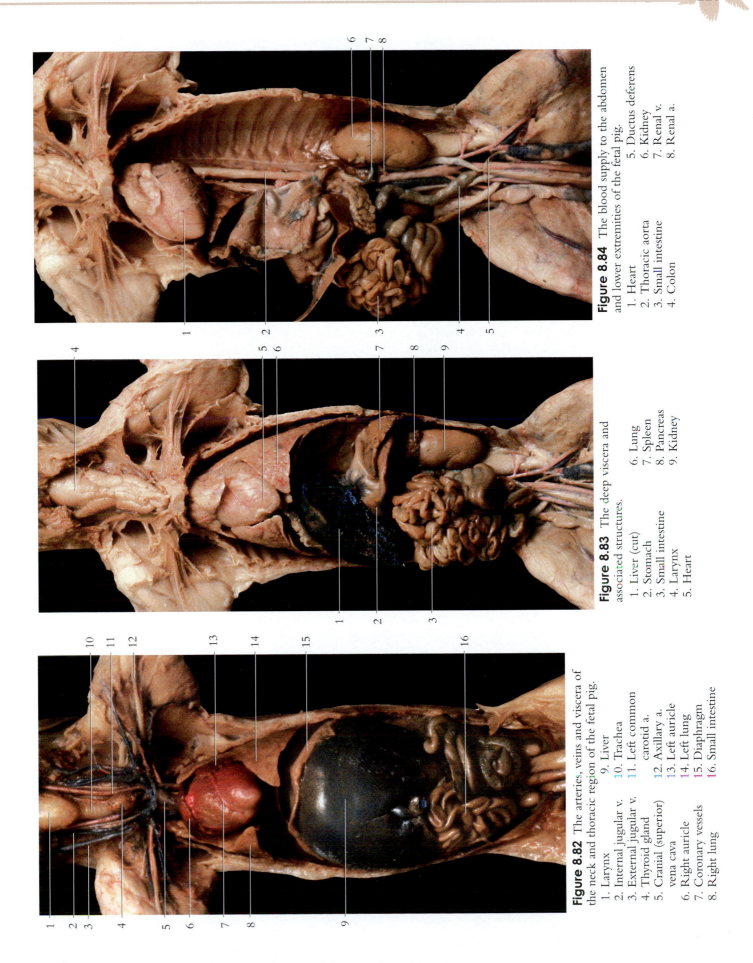

Figure 8.82 The arteries, veins and viscera of the neck and thoracic region of the fetal pig.

1. Larynx
2. Internal jugular v.
3. External jugular v.
4. Thyroid gland
5. Cranial (superior) vena cava
6. Right auricle
7. Coronary vessels
8. Right lung
9. Liver
10. Trachea
11. Left common carotid a.
12. Axillary a.
13. Left auricle
14. Left lung
15. Diaphragm
16. Small intestine

Figure 8.83 The deep viscera and associated structures.

1. Liver (cut)
2. Stomach
3. Small intestine
4. Larynx
5. Heart
6. Lung
7. Spleen
8. Pancreas
9. Kidney

Figure 8.84 The blood supply to the abdomen and lower extremities of the fetal pig.

1. Heart
2. Thoracic aorta
3. Small intestine
4. Colon
5. Ductus deferens
6. Kidney
7. Renal v.
8. Renal a.

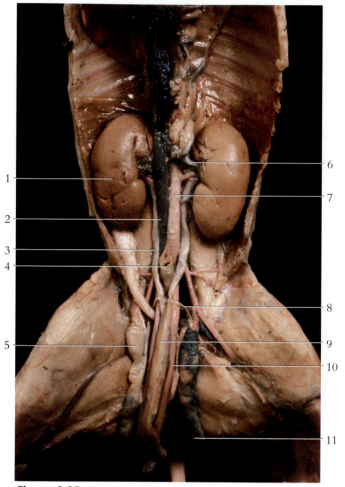

Figure 8.85 The urogenital system of the fetal pig.
1. Right kidney
2. Caudal (inferior) vena cava
3. Ureter
4. Rectum (cut)
5. Partially descended testis
6. Renal vein
7. Descending aorta
8. Vas (ductus) deferens
9. Urinary bladder
10. Umbilical artery
11. Epididymis

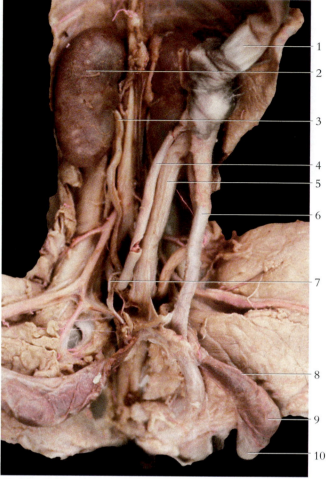

Figure 8.86 The urogenital system of the fetal pig.
1. Umbilical cord
2. Right kidney
3. Ureter
4. Umbilical artery
5. Urinary bladder
6. Penis
7. Vas (ductus) deferens
8. Spermatic cord
9. Left testis
10. Epididymis

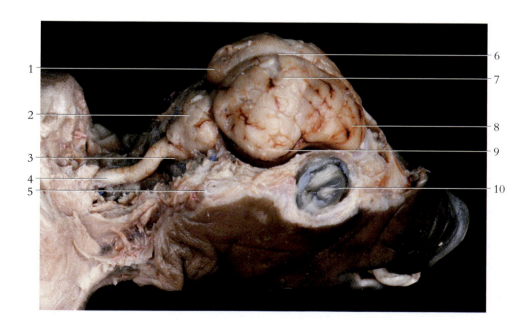

Figure 8.87 The general structures of the fetal pig brain. Because the cerebrum is less defined in pigs, the regions are not known as lobes as they are in humans.
1. Occipital region of cerebrum
2. Cerebellum
3. Medulla oblongata
4. Spinal cord
5. External acoustic meatus
6. Longitudinal fissure
7. Parietal region of cerebrum
8. Frontal region of cerebrum
9. Temporal region of cerebrum
10. Eye

Cat Dissection

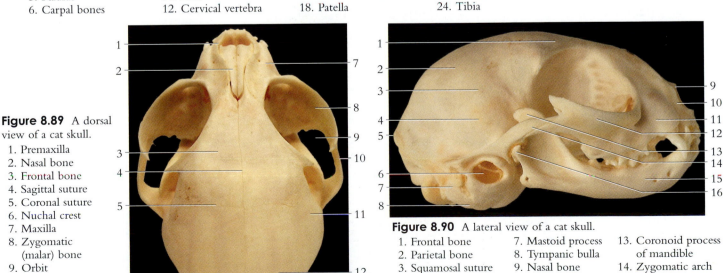

Figure 8.88 The cat skeleton.

1. Mandible
2. Hyoid bone
3. Humerus
4. Ulna
5. Radius
6. Carpal bones

7. Metacarpal bones
8. Phalanges
9. Skull
10. Atlas
11. Axis
12. Cervical vertebra

13. Scapula
14. Sternum
15. Rib
16. Thoracic vertebra
17. Lumbar vertebra
18. Patella

19. Ilium
20. Ischium
21. Pubis
22. Caudal vertebra
23. Femur
24. Tibia

25. Fibula
26. Tarsal bones
27. Metatarsal bones
28. Phalanges

Figure 8.89 A dorsal view of a cat skull.

1. Premaxilla
2. Nasal bone
3. Frontal bone
4. Sagittal suture
5. Coronal suture
6. Nuchal crest
7. Maxilla
8. Zygomatic (malar) bone
9. Orbit
10. Zygomatic arch
11. Temporal bone
12. Parietal bone
13. Interparietal bone

Figure 8.90 A lateral view of a cat skull.

1. Frontal bone
2. Parietal bone
3. Squamosal suture
4. Temporal bone
5. Nuchal crest
6. External acoustic meatus

7. Mastoid process
8. Tympanic bulla
9. Nasal bone
10. Premaxilla bone
11. Maxilla
12. Zygomatic (malar) bone

13. Coronoid process of mandible
14. Zygomatic arch
15. Mandible
16. Condylar process of mandible

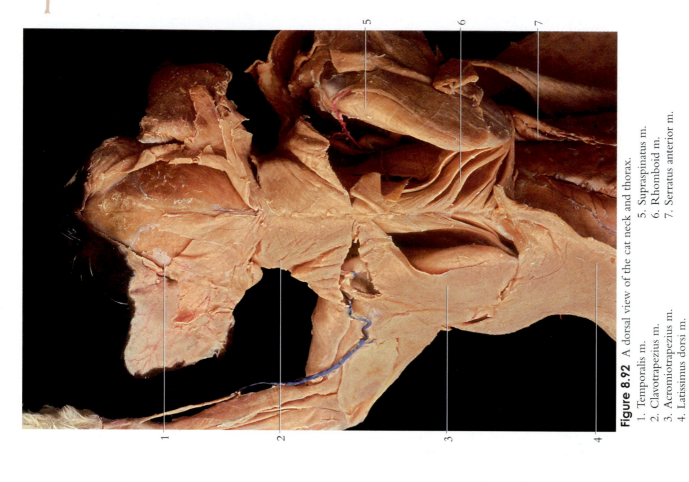

Figure 8.92 A dorsal view of the cat neck and thorax.

1. Temporalis m.
2. Clavotrapezius m.
3. Acromiotrapezius m.
4. Latissimus dorsi m.
5. Supraspinatus m.
6. Rhomboid m.
7. Serratus anterior m.

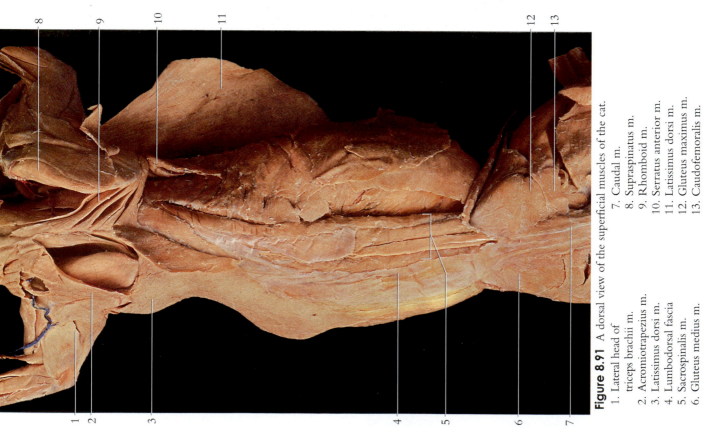

Figure 8.91 A dorsal view of the superficial muscles of the cat.

1. Lateral head of
   triceps brachii m.
2. Acromiotrapezius m.
3. Latissimus dorsi m.
4. Lumbodorsal fascia
5. Sacrospinalis m.
6. Gluteus medius m.
7. Caudal m.
8. Supraspinatus m.
9. Rhomboid m.
10. Serratus anterior m.
11. Latissimus dorsi m.
12. Gluteus maximus m.
13. Caudofemoralis m.

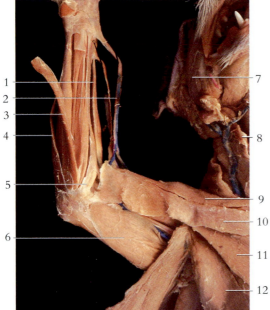

Figure 8.93 An anterior view of the cat brachium and antebrachium.

1. Extensor carpi radialis longus m.
2. Brachioradialis m.
3. Palmaris longus m. (cut)
4. Flexor carpi ulnaris m.
5. Pronator teres m.
6. Epitrochlearis
7. Masseter m.
8. Sternomastoid m.
9. Clavobrachialis m.
10. Pectoantebrachialis m.
11. Pectoralis major m.
12. Pectoralis minor m.

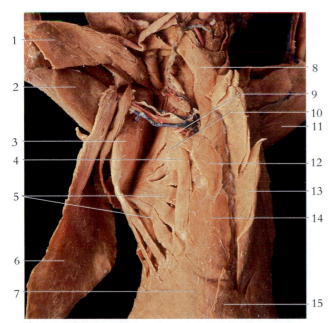

Figure 8.95 An anterior view of the cat trunk.

1. Pectoralis minor (cut)
2. Epitrochlearis m.
3. Subscapularis m.
4. Scalenus medius m.
5. Serratus anterior m.
6. Latissimus dorsi m. (cut)
7. External abdominal oblique m.
8. Sternomastoid m.
9. Scalenus anterior m.
10. Scalenus posterior m.
11. Epitrochlearis m.
12. Transverse costarum m.
13. Pectoralis minor m. (cut)
14. Rectus abdominis m.
15. Xiphihumeralis m. (cut)

Figure 8.94 A ventral view of the cat neck and thorax.

1. Digastric m.
2. Mylohyoid m.
3. Sternomastoid m.
4. Clavotrapezius m.
5. Masseter m.
6. Clavobrachialis m.
7. Pectoantebrachialis m.
8. Pectoralis major m.
9. Pectoralis minor m.

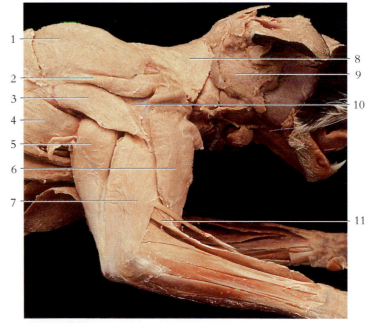

Figure 8.96 A lateral view of the cat shoulder and brachium.

1. Acromiotrapezius m.
2. Levator scapulae ventralis m.
3. Spinodeltoid m.
4. Latissimus dorsi m.
5. Long head of triceps brachii m.
6. Clavobrachialis m.
7. Lateral head of triceps brachii m.
8. Clavotrapezius m.
9. Parotid gland
10. Acromiodeltoid m.
11. Brachioradialis m.

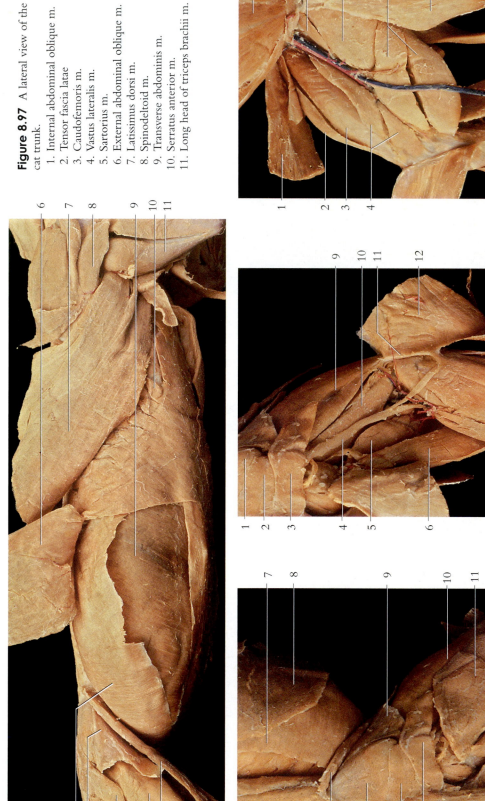

Figure 8.97 A lateral view of the cat trunk.

1. Internal abdominal oblique m.
2. Tensor fascia latae
3. Caudofemoris m.
4. Vastus lateralis m.
5. Sartorius m.
6. External abdominal oblique m.
7. Latissimus dorsi m.
8. Spinodeltoid m.
9. Transverse abdominis m.
10. Serratus anterior m.
11. Long head of triceps brachii m.

Figure 8.98 A lateral view of the cat superficial thigh.

1. Sartorius m.
2. Gluteus medius m.
3. Gluteus maximus m.
4. Caudofemoris m.
5. Caudal m.
6. Semitendinosus m.
7. Internal abdominal oblique m.
8. External abdominal oblique m.
9. Tensor fascia latae (cut)
10. Vastus lateralis m.
11. Biceps femoris m.

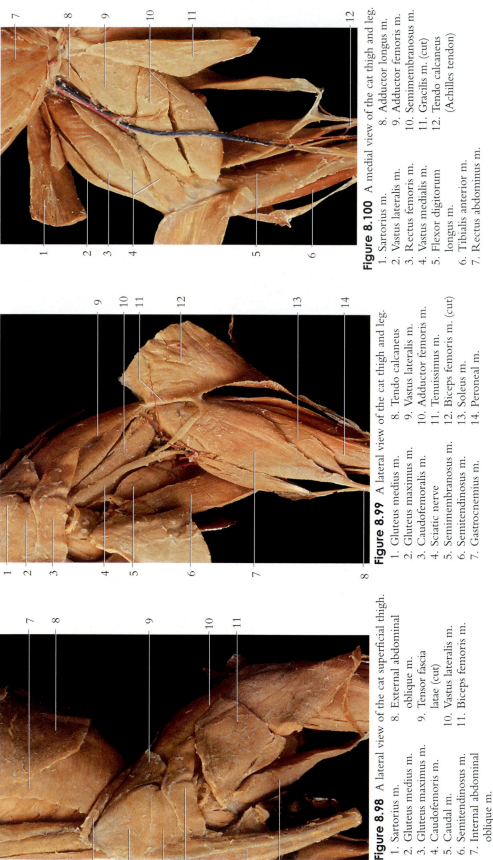

Figure 8.99 A lateral view of the cat thigh and leg.

1. Gluteus medius m.
2. Gluteus maximus m.
3. Caudofemoralis m.
4. Sciatic nerve
5. Semimembranosus m.
6. Semitendinosus m.
7. Gastrocnemius m.
8. Tendo calcaneus
9. Vastus lateralis m.
10. Adductor femoris m.
11. Tenuissimus m.
12. Biceps femoris m. (cut)
13. Soleus m.
14. Peroneal m.

Figure 8.100 A medial view of the cat thigh and leg.

1. Sartorius m.
2. Vastus lateralis m.
3. Rectus femoris m.
4. Vastus medialis m.
5. Flexor digitorum longus m.
6. Tibialis anterior m.
7. Rectus abdominus m.
8. Adductor longus m.
9. Adductor femoris m.
10. Semimembranosus m.
11. Gracilis m. (cut)
12. Tendo calcaneus (Achilles tendon)

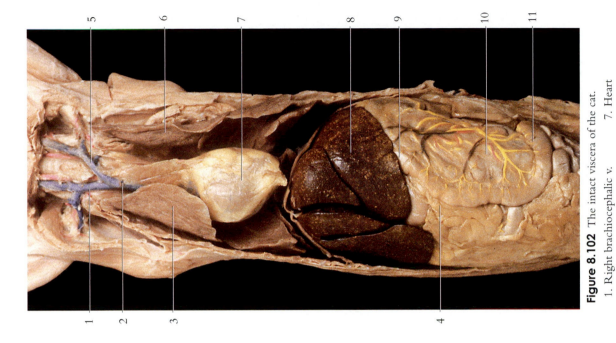

Figure 8.102 The intact viscera of the cat.

1. Right brachiocephalic v.
2. Superior vena cava
3. Right lung
4. Greater omentum
5. Left brachiocephalic v.
6. Left lung
7. Heart
8. Liver
9. Stomach
10. Mesentery
11. Small intestine

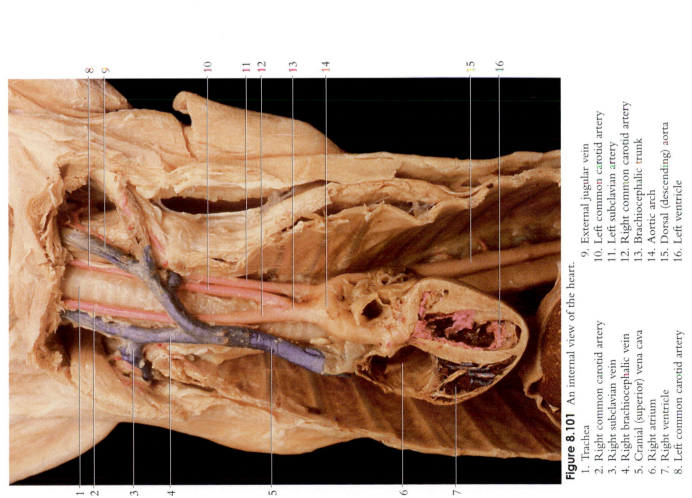

Figure 8.101 An internal view of the heart.

1. Trachea
2. Right common carotid artery
3. Right subclavian vein
4. Right brachiocephalic vein
5. Cranial (superior) vena cava
6. Right atrium
7. Right ventricle
8. Left common carotid artery
9. External jugular vein
10. Left common carotid artery
11. Left subclavian artery
12. Right common carotid artery
13. Brachiocephalic trunk
14. Aortic arch
15. Dorsal (descending) aorta
16. Left ventricle

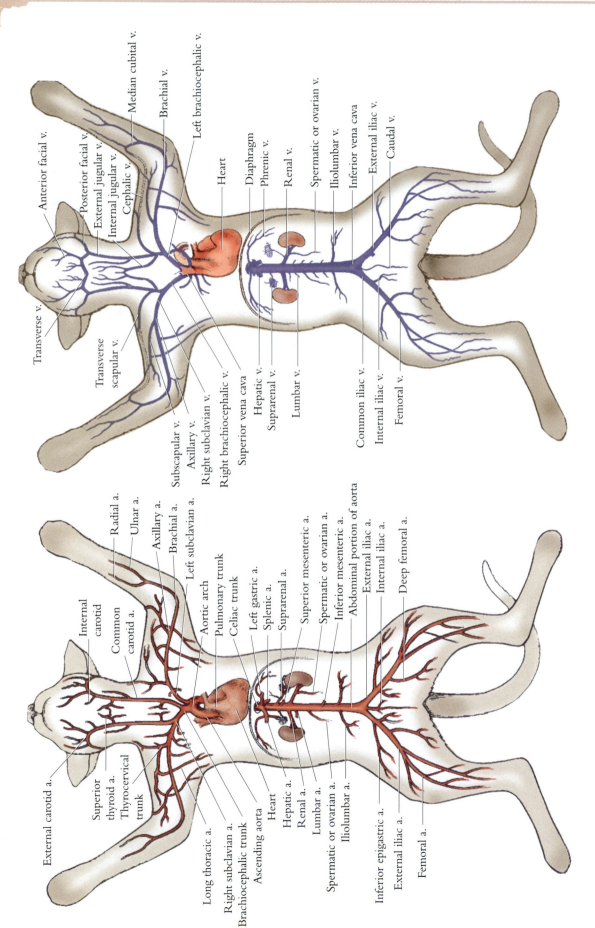

Figure 8.104 The principal veins of the cat, ventral view. (v=vein)

Figure 8.103 The principal arteries of the cat, ventral view. (a=artery)

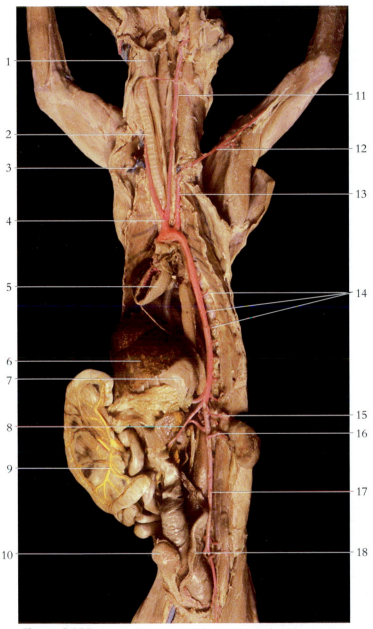

Figure 8.105 An anterior view of the arteries and veins of the trunk of a cat.

1. Larynx
2. Right subclavian vein
3. Trachea
4. Brachiocephalic trunk
5. Heart (cut)
6. Liver
7. Stomach
8. Superior mesenteric artery
9. Superior mesenteric vein
10. Urinary bladder
11. Left common carotid artery
12. Left axillary artery
13. Left subclavian artery
14. Intercostal arteries
15. Adrenolumbar artery
16. Renal artery
17. Abdominal aorta
18. Left horn of uterus

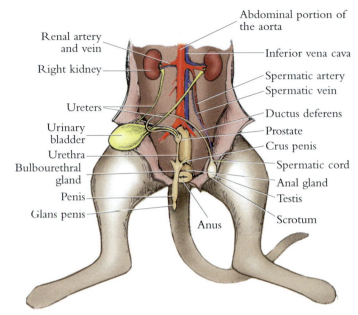

Figure 8.106 A diagram of the urogenital system of a male cat.

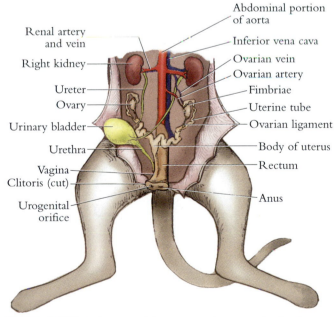

Figure 8.107 A diagram of the urogenital system of a female cat.

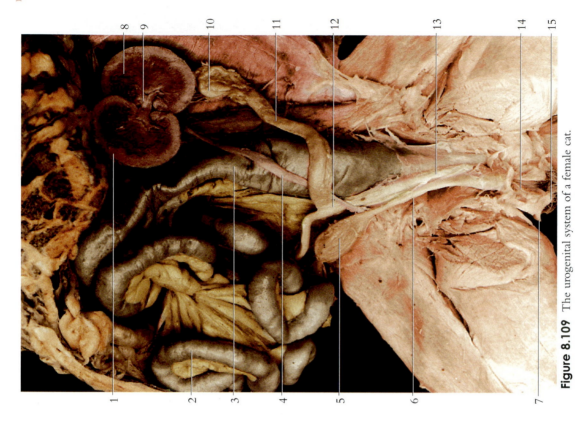

Figure 8.109 The urogenital system of a female cat.

1. Renal cortex	9. Renal pelvis
2. Small intestine	10. Ovary
3. Colon	11. Horn of uterus
4. Ureter	12. Body of uterus
5. Urinary bladder	13. Vagina
6. Urethra	14. Vestibule (split)
7. Clitoris	15. Labia
8. Renal medulla	

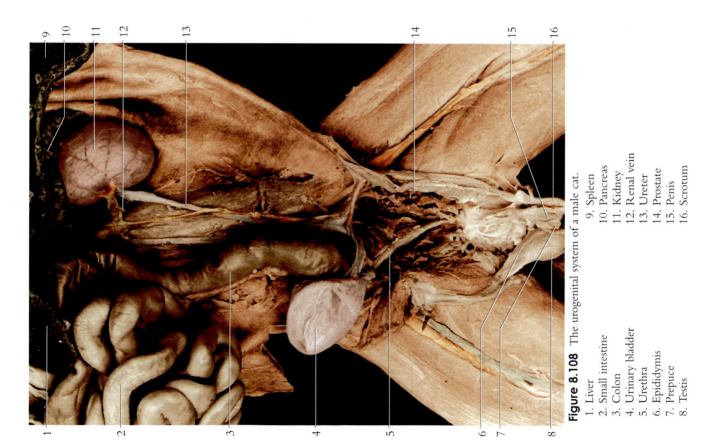

Figure 8.108 The urogenital system of a male cat.

1. Liver	9. Spleen
2. Small intestine	10. Pancreas
3. Colon	11. Kidney
4. Urinary bladder	12. Renal vein
5. Urethra	13. Ureter
6. Epididymis	14. Prostate
7. Prepuce	15. Penis
8. Testis	16. Scrotum

Mammalian Heart and Brain Dissection

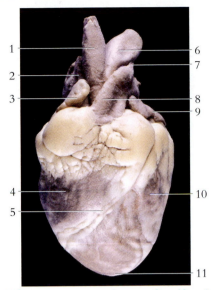

Figure 8.110 A ventral view of mammalian (sheep) heart.
1. Brachiocephalic artery
2. Cranial vena cava
3. Right auricle of right atrium
4. Right ventricle
5. Interventicular groove
6. Aortic arch
7. Ligamentum arteriosum
8. Pulmonary trunk
9. Left auricle of left atrium
10. Left ventricle
11. Apex of heart

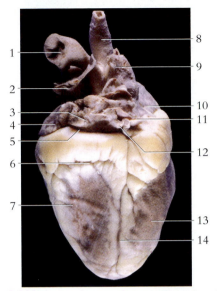

Figure 8.111 A dorsal view of mammalian (sheep) heart.
1. Aorta
2. Pulmonary artery
3. Pulmonary vein
4. Left auricle
5. Left atrium
6. Atrioventricular groove
7. Left ventricle
8. Brachiocephalic artery
9. Cranial vena cava
10. Right auricle
11. Right atrium
12. Pulmonary vein
13. Right ventricle
14. Interventricular groove

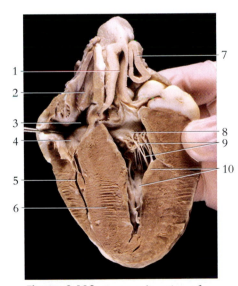

Figure 8.112 A coronal section of the mammalian (sheep) heart.
1. Aorta
2. Cranial vena cava
3. Right atrium
4. Right atrioventricular (tricuspid) valve
5. Right ventricle
6. Interventricular septum
7. Pulmonary artery
8. Left atrioventricular (bicuspid) valve
9. Chordae tendineae
10. Papillary muscles

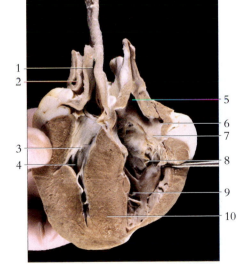

Figure 8.113 A coronal section of the mammalian (sheep) heart showing the valves.
1. Opening of the brachiocephalic artery
2. Pulmonary artery
3. Left atrioventricular (bicuspid) valve
4. Left ventricle
5. Opening of cranial vena cava
6. Opening of coronary sinus
7. Right atrium
8. Right atrioventricular (tricuspid) valve
9. Right ventricle
10. Interventricular septum

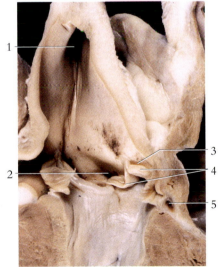

Figure 8.114 A coronal section of the mammalian (sheep) heart showing openings of coronary arteries.
1. Opening of brachiocephalic artery
2. Opening of left coronary artery
3. Opening of right coronary artery
4. Aortic valve
5. Coronary vessel

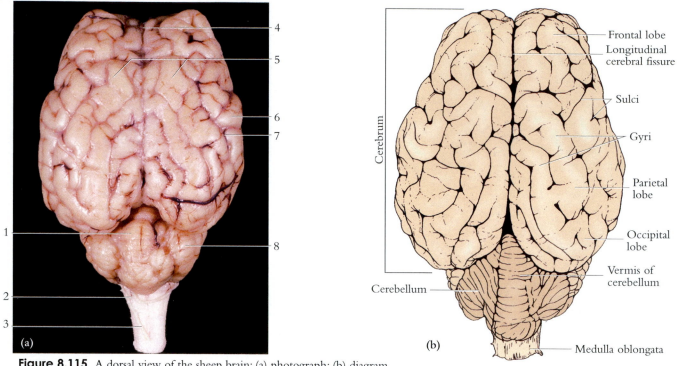

Figure 8.115 A dorsal view of the sheep brain: (a) photograph; (b) diagram.

1. Vermis
2. Medulla oblongata
3. Spinal cord
4. Longitudinal cerebral fissure
5. Cerebral hemispheres
6. Gyrus
7. Sulcus
8. Cerebellar hemisphere

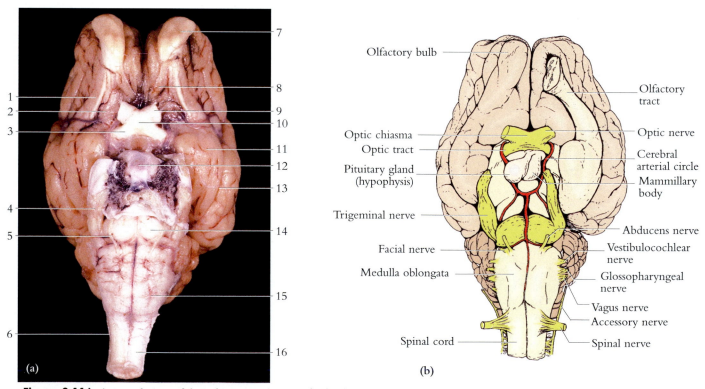

Figure 8.116 A ventral view of sheep brain: (a) photograph; (b) diagram.

1. Lateral olfactory band
2. Olfactory trigone
3. Optic tract
4. Trigeminal nerve
5. Abducens nerve
6. Accessory nerve
7. Olfactory bulb
8. Medial olfactory band
9. Optic nerve
10. Optic chiasma
11. Pyriform lobe
12. Pituitary gland (hypophysis)
13. Rhinal sulcus
14. Pons
15. Medulla oblongata
16. Spinal cord

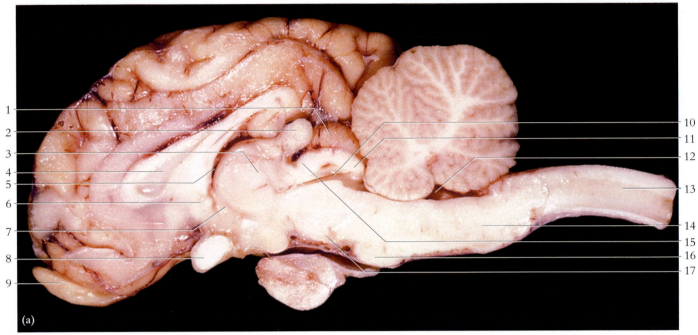

Figure 8.117 A right sagittal view of the sheep brain: (a) photograph; (b) diagram.

1. Superior colliculus
2. Pineal body (gland)
3. Intermediate mass of thalamus
4. Septum pellucidum
5. Interventicular foramen (foramen of Monro)
6. Anterior commissure
7. Third ventricle
8. Optic chiasma
9. Olfactory bulb
10. Mesencephalic (cerebral) aqueduct
11. Inferior colliculus
12. Fourth ventricle
13. Spinal cord
14. Medulla oblongata
15. Posterior commissure
16. Pons
17. Cerebral peduncle

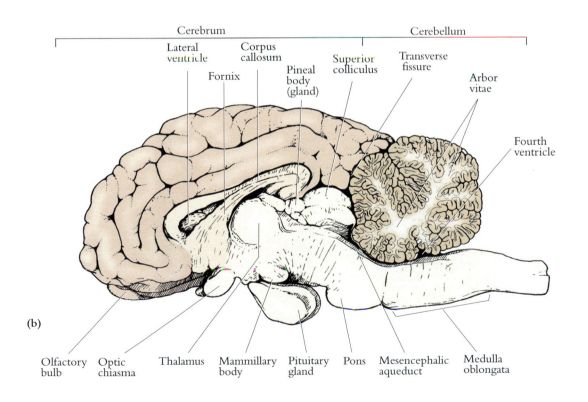

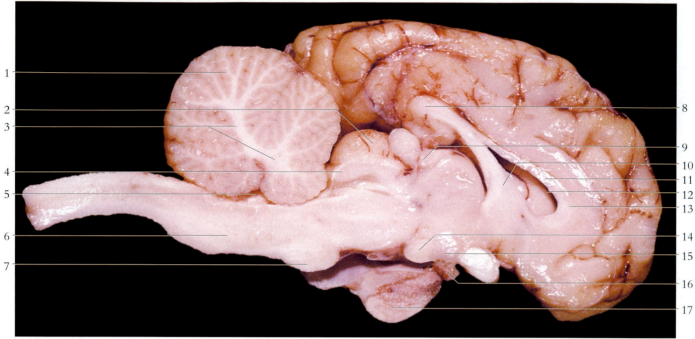

Figure 8.118 A left sagittal view of the sheep brain.

1. Cerebellum
2. Superior colliculus
3. Arbor vitae
4. Inferior colliculus
5. Fourth ventricle
6. Medulla oblongata

7. Pons
8. Splenium of corpus callosum
9. Habenular trigone
10. Fornix
11. Body of corpus callosum
12. Lateral ventricle

13. Genu of corpus callosum
14. Mammillary body
15. Tuber cinereum of hypothalamus
16. Pituitary stalk
17. Pituitary gland (hypophysis)

Figure 8.119 A lateral view of the brainstem.

1. Pons
2. Abducens nerve
3. Medulla oblongata
4. Hypoglossal nerve
5. Spinal cord
6. Lateral geniculate body
7. Medial geniculate body
8. Trochlear nerve
9. Trigeminal nerve
10. Accessory nerve

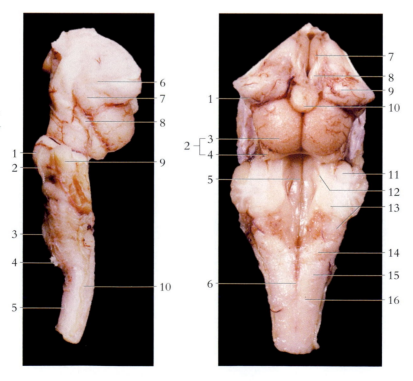

Figure 8.120 A dorsal view of the brainstem.

1. Medial geniculate body
2. Corpora quadrigemina
3. Superior colliculus
4. Inferior colliculus
5. Fourth ventricle
6. Dorsal median sulcus
7. Intermediate mass of thalamus
8. Habenular trigone
9. Thalamus
10. Pineal gland
11. Middle cerebellar peduncle
12. Anterior cerebellar penduncle
13. Posterior cerebellar penduncle
14. Tuberculum cuneatum
15. Fasciculus gracilis
16. Fasciculus cuneatus

Because humans are vertebrate organisms, the study of human biology is appropriate in a general biology course. *Human anatomy* is the scientific discipline that investigates the structure of the body, and *human physiology* is the scientific discipline that investigates how body structures function. The purpose of this chapter is to present a visual overview of the principal anatomical structures of the human body.

Since both the *skeletal system* and the *muscular system* are concerned with body movement, they are frequently discussed together as the *skeletomusculature system*. In a functional sense, the flexible internal framework, or *bones* of the skeleton, support and provide movement at the *joints* where the muscles attached to the bones produce their actions as they are stimulated to contract.

The *nervous system* is anatomically divided into the *central nervous system* (CNS), which includes the *brain* and *spinal cord*, and the *peripheral nervous system* (PNS), which includes the *cranial nerves*, arising from the brain, and the *spinal nerves*, arising from the spinal cord. The *autonomic nervous system* (ANS) is a functional division of the nervous system devoted to regulation of involuntary activities of the body. The brain and spinal cord are the centers for integration and coordination of information. *Nerves*, composed of *neurons*, convey nerve impulses to and from the brain. *Sensory organs*, such as the eyes and ears, respond to impulses in the environment and convey sensations to the CNS. The nervous system functions with the *endocrine system* in coordinating body activities.

The *cardiovascular system* consists of the *heart, vessels* (both blood and lymphatic vessels), *blood*, and the *tissues* that produce the *blood*. The four-chambered human heart is enclosed with a *pericardial sac* within the thoracic cavity. *Arteries* and *arterioles* transport blood away from the heart, *capillaries* permeate the tissues and are the functional units for product exchange with the cells, and *venules* and *veins* transport blood toward the heart. *Lymphatic vessels* return interstitial fluid back to the circulatory system after first passing it through *lymph nodes* for cleansing. Blood cells are produced in the bone marrow and once old and worn, they are broken down in the liver.

The *respiratory system* consists of the *conducting division* that transports air to and from the *respiratory division* within the *lungs*. The *alveoli* of the lungs contact the capillaries of the cardiovascular system and are the sites for transport of respiratory gases into and out of the body.

The *digestive system* consists of a *gastrointestinal tract* (GI tract) and *accessory digestive organs*. Food traveling through the GI tract is processed such that it is suitable for absorption through the intestinal wall into the bloodstream. The *pancreas* and *liver* are the principal digestive organs. The pancreas produces hormones and enzymes. The liver processes nutrients, stores glucose as glycogen, and excretes bile.

Because of commonality of prenatal development and dual functions of some of the organs, the *urinary system* and *reproductive*

system may be considered together as the *urogenital system*. The urinary system, consisting of the *kidneys, ureters, urinary bladder*, and *urethra*, extracts and processes wastes from the blood in the form of urine. The male and female reproductive systems produce regulatory hormones and gametes (sperm and ova, respectively) within the gonads (testes and ovaries). Sexual reproduction is the mechanism for propagation of offspring that have traits from both parents. The process of prenatal development is made possible by the formation of *extraembryonic membranes* (placenta, umbilical cord, allantois, amnion, chorion, and yolk sac) within the uterus of the pregnant woman.

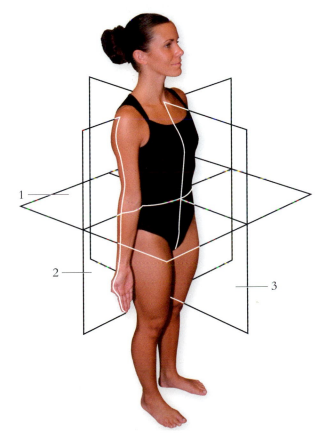

Figure 9.1 The planes of reference in a person while standing in anatomical position. The anatomical position provides a basis of reference for describing the relationship of one body part to another. In the anatomical position, the person is standing, the feet are parallel, the eyes are directed forward, and the arms are to the sides with the palms turned forward and the fingers pointed straight down.

1. Transverse plane
   (cross-sectional plane)
2. Coronal plane
   (frontal plane)
3. Sagittal plane

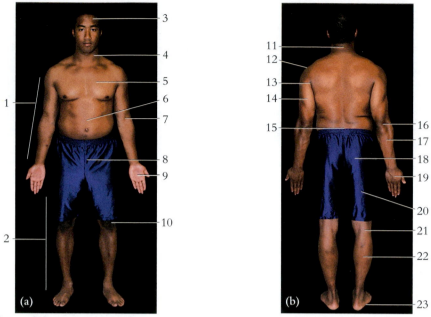

Figure 9.2 The major body parts and regions in humans (bipedal vertebrate). (a) An anterior view, and (b) a posterior view.

1. Upper extremity
2. Lower extremity
3. Head
4. Neck, anterior aspect
5. Thorax (chest)
6. Abdomen
7. Cubital fossa
8. Pubic region

9. Palmar region (palm)
10. Patellar region (patella)
11. Cervical region
12. Shoulder
13. Axilla (armpit)
14. Brachium (upper arm)
15. Lumbar region
16. Elbow

17. Antebrachium (forearm)
18. Gluteal region (buttock)
19. Dorsum of hand
20. Thigh
21. Popliteal fossa
22. Calf
23. Plantar surface (sole)

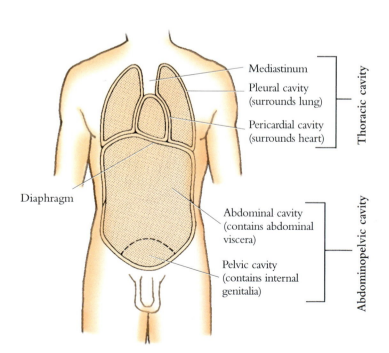

Figure 9.3 An anterior view (coronal plane) of the body cavities of the trunk.

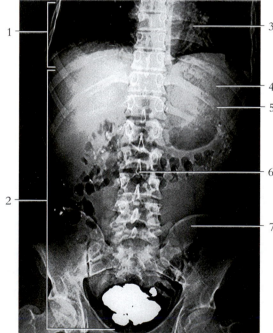

Figure 9.4 An MR image of the trunk showing the body cavities and their contents.

1. Thoracic cavity
2. Abdominopelvic cavity
3. Image of heart
4. Image of diaphragm

5. Image of rib
6. Image of lumbar vertebra
7. Image of ilium

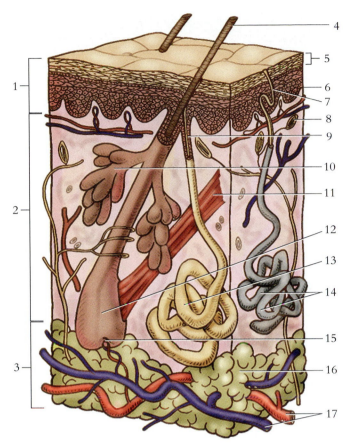

Figure 9.5 The skin and certain epidermal structures.

1. Epidermis
2. Dermis
3. Hypodermis
4. Shaft of hair
5. Stratum corneum
6. Stratum basale
7. Sweat duct
8. Sensory receptor
9. Sweat duct
10. Sebaceous gland
11. Arrector pili muscle
12. Hair follicle
13. Apocrine sweat gland
14. Eccrine sweat gland
15. Bulb of hair
16. Adipose tissue
17. Cutaneous blood vessels

Figure 9.6 The gross structure of the skin and underlying fascia.

1. Epidermis
2. Dermis
3. Hypodermis
4. Fascia
5. Muscle

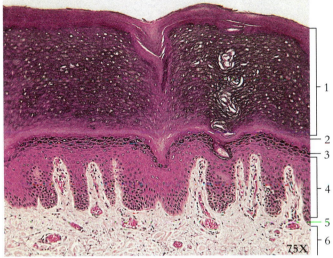

Figure 9.7 The epidermis and dermis of thick skin.

1. Stratum corneum
2. Stratum lucidum
3. Stratum granulosum
4. Stratum spinosum
5. Stratum basale
6. Dermis

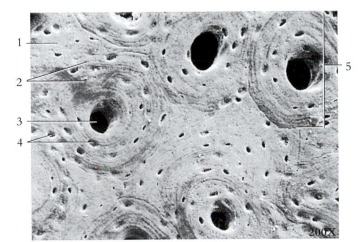

Figure 9.8 An electron micrograph of bone tissue.

1. Interstitial lamellae
2. Lamellae
3. Central canal (Haversian canal)
4. Lacunae
5. Osteon (Haversian system)

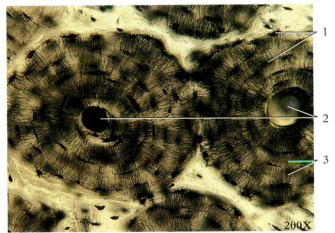

Figure 9.9 A transverse section of two osteons.

1. Lacunae with contained osteocytes
2. Central (Haversian) canals
3. Lamellae

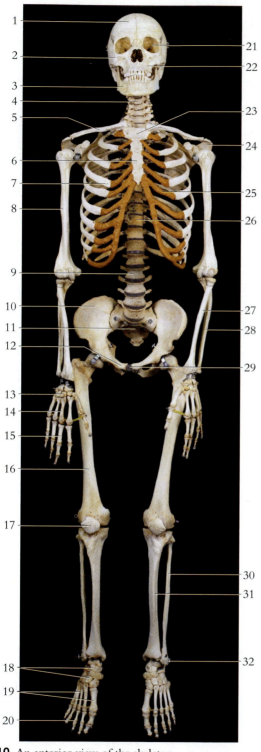

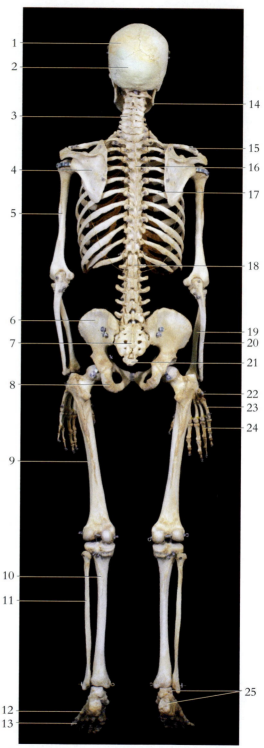

Figure 9.10 An anterior view of the skeleton.

1. Frontal bone	12. Pubis	23. Manubrium
2. Zygomatic bone	13. Carpal bones	24. Scapula
3. Mandible	14. Metacarpal bones	25. Costal cartilage
4. Cervical vertebra	15. Phalanges	26. Thoracic vertebra
5. Clavicle	16. Femur	27. Radius
6. Body of sternum	17. Patella	28. Ulna
7. Rib	18. Tarsal bones	29. Symphysis pubis
8. Humerus	19. Metatarsal bones	30. Fibula
9. Lumbar vertebra	20. Phalanges	31. Tibia
10. Ilium	21. Orbit	32. Lateral malleolus
11. Sacrum	22. Maxilla	

Figure 9.11 A posterior view of the skeleton.

1. Parietal bone	10. Tibia	19. Radius
2. Occipital bone	11. Fibula	20. Ulna
3. Cervical vertebra	12. Metatarsal bones	21. Coccyx
4. Scapula	13. Phalanges	22. Carpal bones
5. Humerus	14. Mandible	23. Metacarpal bones
6. Ilium	15. Clavicle	24. Phalanges
7. Sacrum	16. Thoracic vertebra	25. Tarsal bones
8. Ischium	17. Rib	
9. Femur	18. Lumbar vertebra	

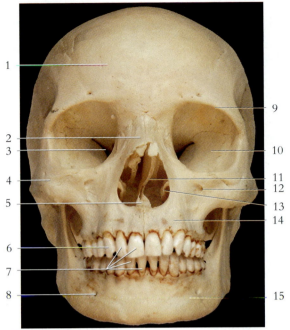

Figure 9.12 An anterior view of the human skull.

1. Frontal bone
2. Nasal bone
3. Superior orbital fissure
4. Zygomatic bone
5. Vomer
6. Canine
7. Incisors
8. Mental foramen
9. Supraorbital margin
10. Sphenoid bone
11. Perpendicular plate of ethmoid bone
12. Infraorbital foramen
13. Inferior nasal concha
14. Maxilla
15. Mandible

Figure 9.13 A lateral view of the human skull.

1. Coronal suture
2. Frontal bone
3. Lacrimal bone
4. Nasal bone
5. Zygomatic bone
6. Maxilla
7. Premolars
8. Molars
9. Mandible
10. Parietal bone
11. Squamosal suture
12. Temporal bone
13. Lambdoidal suture
14. External acoustic meatus
15. Occipital bone
16. Condylar process of mandible
17. Mandibular notch
18. Mastoid process of temporal bone
19. Coronoid process of mandible
20. Angle of mandible

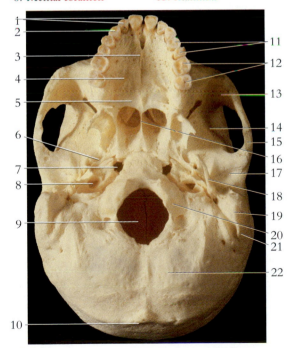

Figure 9.14 An inferior view of the human skull.

1. Incisors
2. Canine
3. Intermaxillary suture
4. Maxilla
5. Palatine bone
6. Foramen ovale
7. Foramen lacerum
8. Carotid canal
9. Foramen magnum
10. Superior nuchal line
11. Premolars
12. Molars
13. Zygomatic bone
14. Sphenoid bone
15. Zygomatic arch
16. Vomer
17. Mandibular fossa
18. Styloid process of temporal bone
19. Mastoid process of temporal bone
20. Occipital condyle
21. Temporal bone
22. Occipital bone

Figure 9.15 A sagittal view of the human skull.

1. Frontal bone
2. Frontal sinus
3. Crista galli of ethmoid bone
4. Cribriform plate of ethmoid bone
5. Nasal bone
6. Nasal concha
7. Maxilla
8. Mandible
9. Parietal bone
10. Occipital bone
11. Internal acoustic meatus
12. Sella turcica
13. Hypoglossal canal
14. Sphenoidal sinus
15. Styloid process of temporal bone
16. Vomer

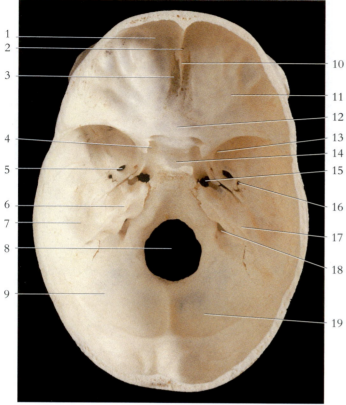

Figure 9.16 A superior view of the cranium.

1. Frontal bone
2. Foramen cecum
3. Cribriform plate of ethmoid bone
4. Optic canal
5. Foramen ovale
6. Petrous part of temporal bone
7. Temporal bone
8. Foramen magnum
9. Occipital bone
10. Crista galli of ethmoid bone
11. Anterior cranial fossa
12. Sphenoid bone
13. Foramen rotundum
14. Sella turcica of sphenoid bone
15. Foramen lacerum
16. Foramen spinosum
17. Internal acoustic meatus
18. Jugular foramen
19. Posterior cranial fossa

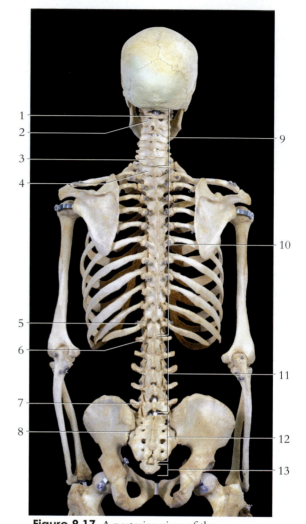

Figure 9.17 A posterior view of the vertebral column.

1. Atlas
2. Axis
3. Seventh cervical vertebra
4. First thoracic vertebra
5. Twelfth thoracic vertebra
6. First lumbar vertebra
7. Fifth lumbar vertebra
8. Sacroiliac joint
9. Cervical vertebrae
10. Thoracic vertebrae
11. Lumbar vertebrae
12. Sacrum
13. Coccyx

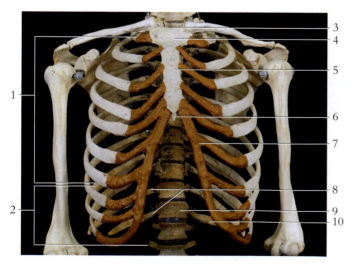

Figure 9.18 An anterior view of the rib cage.

1. True ribs (seven pairs)
2. False ribs (five pairs)
3. Jugular notch
4. Manubrium
5. Body of sternum
6. Xiphoid process
7. Costal cartilage
8. Floating ribs (inferior two pairs of false ribs)
9. Twelfth thoracic vertebra
10. Twelfth rib

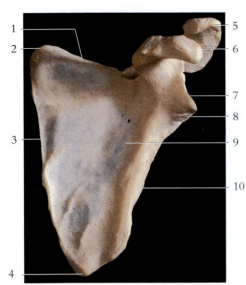

Figure 9.19 An anterior view of the left scapula.
1. Superior border
2. Superior angle
3. Medial (vertebral) border
4. Inferior angle
5. Acromion
6. Coracoid process
7. Glenoid fossa
8. Infraglenoid tubercle
9. Subscapular fossa
10. Lateral (axillary) border

Figure 9.20 A posterior view of the left scapula.
1. Acromion
2. Glenoid fossa
3. Lateral (axillary) border
4. Superior angle
5. Supraspinous fossa
6. Spine
7. Infraspinous fossa
8. Medial (vertebral) border
9. Inferior angle

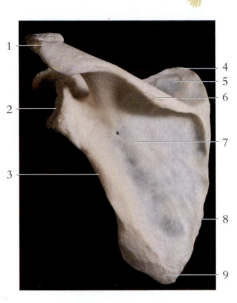

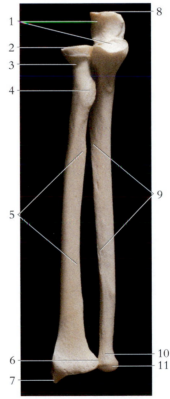

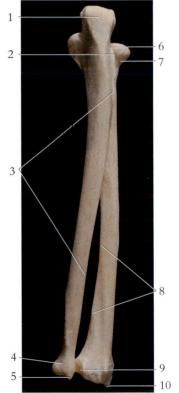

Figure 9.21 The right humerus. (a) Anterior view (b) Posterior view.
1. Greater tubercle
2. Intertubercular groove
3. Lesser tubercle
4. Deltoid tuberosity
5. Anterior body (shaft) of humerus
6. Lateral supracondylar ridge
7. Lateral epicondyle
8. Capitulum
9. Head of humerus
10. Surgical neck
11. Posterior body (shaft) of humerus
12. Olecranon fossa
13. Coronoid fossa
14. Medial epicondyle
15. Trochlea
16. Anatomical neck
17. Greater tubercle
18. Lateral epicondyle

Figure 9.22 An anterior view of the right ulna and radius.
1. Trochlear notch
2. Head of radius
3. Neck of radius
4. Radial tuberosity
5. Interosseous margin
6. Location of ulnar notch of radius
7. Styloid process of radius
8. Olecranon
9. Interosseous margin
10. Neck of ulna
11. Head of ulna

Figure 9.23 A posterior view of the right ulna and radius.
1. Olecranon
2. Location of radial notch of ulna
3. Interosseous margin
4. Head of ulna
5. Styloid process of ulna
6. Head of radius
7. Neck of radius
8. Interosseous margin
9. Ulnar notch
10. Styloid process of radius

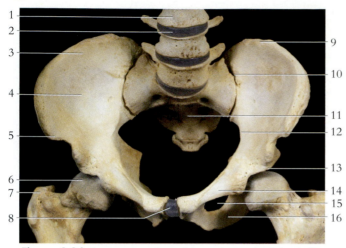

Figure 9.24 An anterior view of the articulated pelvic girdle showing the two coxal bones, the sacrum, and the two femora.

1. Lumbar vertebra
2. Intervertebral disk
3. Ilium
4. Iliac fossa
5. Anterior superior iliac spine
6. Head of femur
7. Greater trochanter
8. Symphysis pubis
9. Crest of the ilium
10. Sacroiliac joint
11. Sacrum
12. Pelvic brim
13. Acetabulum
14. Pubic crest
15. Obturator foramen
16. Ischium

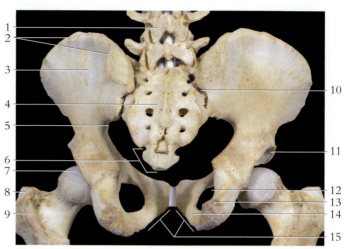

Figure 9.25 A posterior view of the articulated pelvic girdle showing the two coxal bones, the sacrum, and the two femora.

1. Lumbar vertebra
2. Crest of ilium
3. Ilium
4. Sacrum
5. Greater sciatic notch
6. Coccyx
7. Head of femur
8. Greater trochanter
9. Intertrochanteric crest
10. Sacroiliac joint
11. Acetabulum
12. Obturator foramen
13. Ischium
14. Pubis
15. Pubic angle

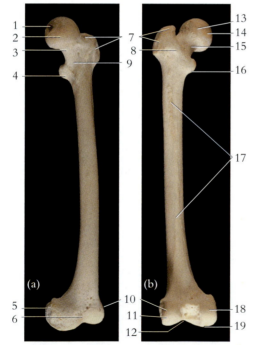

Figure 9.26 The left femur. (a) An anterior view and (b) a posterior view.

1. Fovea capitis femoris
2. Head
3. Neck
4. Lesser trochanter
5. Medial epicondyle
6. Patellar surface
7. Greater trochanter
8. Intertrochanteric crest
9. Intertrochanteric line
10. Lateral epicondyle
11. Lateral condyle
12. Intercondylar fossa
13. Head
14. Fovea capitis femoris
15. Neck
16. Lesser trochanter
17. Linea aspera on shaft (body) of femur
18. Medial epicondyle
19. Medial condyle

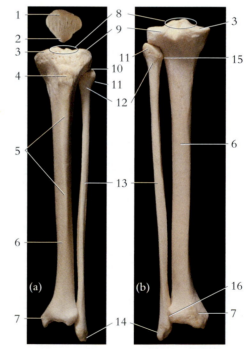

Figure 9.27 (a) An anterior view of the left patella, tibia, and fibula. (b) A posterior view of the left tibia and fibula.

1. Base of patella
2. Apex of patella
3. Medial condyle
4. Tibial tuberosity
5. Anterior crest of tibia
6. Body (shaft) of tibia
7. Medial malleolus
8. Intercondylar tubercles
9. Lateral condyle
10. Tibial articular facet of fibula
11. Head of fibula
12. Neck of fibula
13. Body (shaft) of fibula
14. Lateral malleolus
15. Fibular articular facet of tibia
16. Fibular notch of tibia

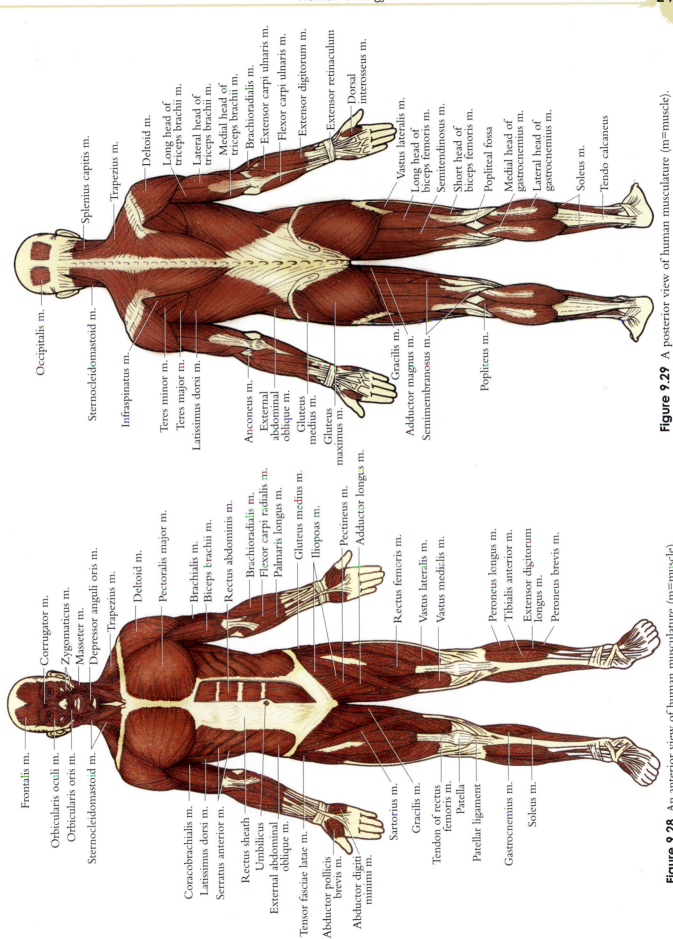

Occipitalis m.

Sternocleidomastoid m.

Infraspinatus m.

Teres minor m.

Teres major m.

Latissimus dorsi m.

Anconeus m.

External
abdominal
oblique m.

Gluteus
medius m.

Gluteus
maximus m.

Gracilis m.

Adductor magnus m.

Semimembranosus m.

Popliteus m.

Splenius capitis m.

Trapezius m.

Deltoid m.

Long head of
triceps brachii m.

Lateral head of
triceps brachii m.

Medial head of
triceps brachii m.

Brachioradialis m.

Extensor carpi ulnaris m.

Flexor carpi ulnaris m.

Extensor digitorum m.

Extensor retinaculum

Dorsal
interosseus m.

Vastus lateralis m.

Long head of
biceps femoris m.

Semitendinosus m.

Short head of
biceps femoris m.

Popliteal fossa

Medial head of
gastrocnemius m.

Lateral head of
gastrocnemius m.

Soleus m.

Tendo calcaneus

Figure 9.29 A posterior view of human musculature (m=muscle).

Frontalis m.

Orbicularis oculi m.

Orbicularis oris m.

Sternocleidomastoid m.

Coracobrachialis m.

Latissimus dorsi m.

Serratus anterior m.

Rectus sheath

Umbilicus

External abdominal
oblique m.

Tensor fasciae latae m.

Abductor pollicis
brevis m.

Abductor digiti
minimi m.

Corrugator m.

Zygomaticus m.

Masseter m.

Depressor anguli oris m.

Trapezius m.

Deltoid m.

Pectoralis major m.

Brachialis m.

Biceps brachii m.

Rectus abdominis m.

Brachioradialis m.

Flexor carpi radialis m.

Palmaris longus m.

Gluteus medius m.

Iliopoas m.

Pectineus m.

Adductor longus m.

Sartorius m.

Gracilis m.

Tendon of rectus
femoris m.

Patella

Patellar ligament

Gastrocnemius m.

Soleus m.

Rectus femoris m.

Vastus lateralis m.

Vastus medialis m.

Peroneus longus m.

Tibialis anterior m.

Extensor digitorum
longus m.

Peroneus brevis m.

Figure 9.28 An anterior view of human musculature (m=muscle).

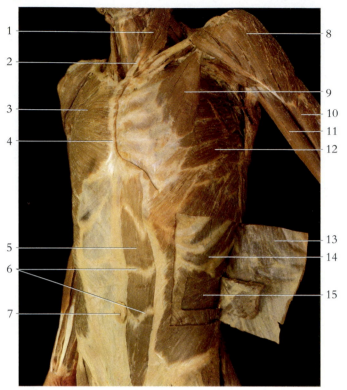

Figure 9.30 An anterolateral view of the trunk and upper arm.

1. Sternocleidomastoid m.
2. Tendon of sternocleidomastoid m.
3. Pectoralis major m.
4. Sternum.
5. Rectus abdominis m.
6. Tendinous inscriptions of rectus abdominis m.
7. Umbilicus
8. Deltoid m.
9. Pectoralis minor m.
10. Brachialis m.
11. Biceps brachii m. (long head)
12. Serratus anterior m.
13. External abdominal oblique m. (reflected)
14. External intercostal m.
15. Transverse abdominis m.

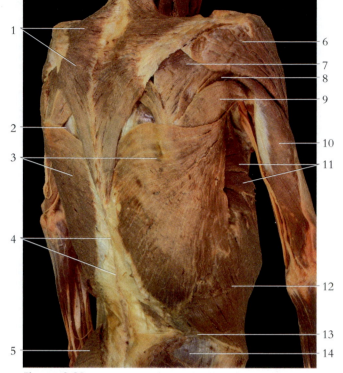

Figure 9.31 A posterolateral view of the trunk and upper arm.

1. Trapezius m.
2. Triangle of ausculation
3. Latissimus dorsi mm.
4. Vertebral column (spinous processes)
5. Gluteus maximus m.
6. Deltoid m.
7. Infraspinatus m.
8. Teres minor m.
9. Teres major m.
10. Triceps brachii m. (long head)
11. Serratus anterior mm.
12. External abdominal oblique m.
13. Iliac crest
14. Gluteus medius m.

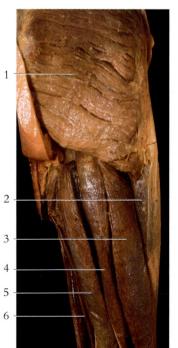

Figure 9.32 The superficial muscles of gluteal and thigh regions.

1. Gluteus maximus m.
2. Vastus lateralis m.
3. Biceps femoris m.
4. Semitendinosus m.
5. Semimembranosus m.
6. Gracilis m.

Figure 9.33 The deep structures of gluteal region.

1. Piriformis m.
2. Sciatic n.
3. Obturator internus m.
4. Quadratus femoris m.
5. Adductor minimus m.
6. Gluteus medius m. (reflected)
7. Gluteus minimus m.
8. Superior gemellus m.
9. Inferior gemellus m.
10. Gluteus maximus m. (reflected)

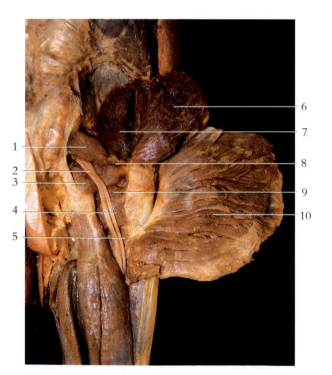

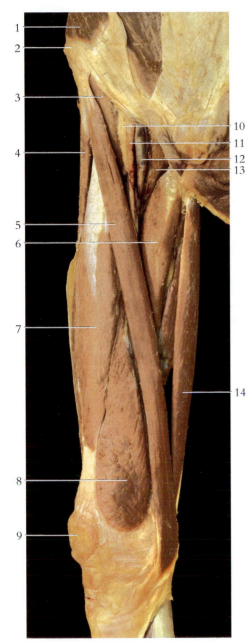

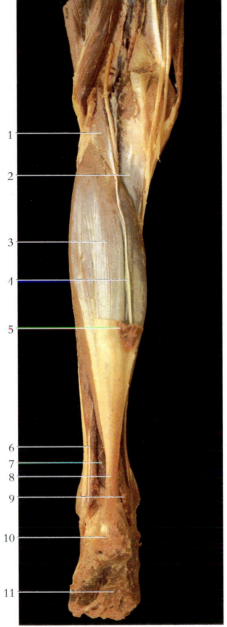

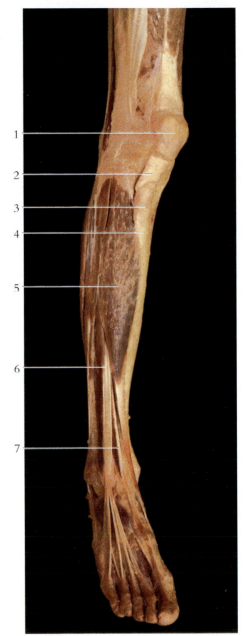

Figure 9.34 An anteromedial view of
the right thigh.
1. External oblique m.
2. Anterior superior iliac spine
3. Iliopsoas m.
4. Tensor fascia lata m.
5. Sartorius m.
6. Adductor longus m.
7. Rectus femoris m.
8. Vastus medialis m.
9. Patella
10. Femoral nerve
11. Femoral artery
12. Femoral vein
13. Pectineus m.
14. Gracilis m.

Figure 9.35 A posterior view of
lower leg.
1. Plantaris m.
2. Popliteus m.
3. Soleus m.
4. Plantaris tendon
5. Gastrocnemius m. (cut)
6. Fibularis longus and brevis mm.
7. Flexor hallucis longus m.
8. Calcaneal tendon (Achilles tendon)
9. Tendon of flexor hallucis longus m.
10. Calcaneus
11. Plantar aponeurosis

Figure 9.36 An anterior view of
lower leg.
1. Patella
2. Patellar ligament
3. Tibial tuberosity
4. Tibia
5. Tibialis anterior m.
6. Tendon of extensor
   digitorum longus m.
7. Tendon of extensor
   hallucis longus m.

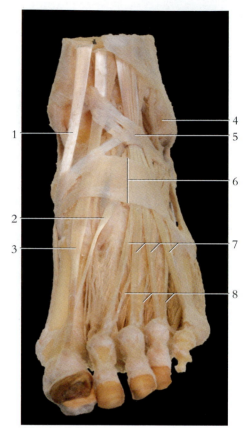

Figure 9.37 An anterior view of dorsum of foot.
1. Tendon of tibialis anterior m.
2. Tendon of extensor hallucis brevis m.
3. Tendon of extensor hallucis longus m.
4. Lateral malleolus
5. Superior extensor retinaculum
6. Inferior extensor retinaculum
7. Tendon of extensor digitorum longus m.
8. Tendon of extensor digitorum brevis m.

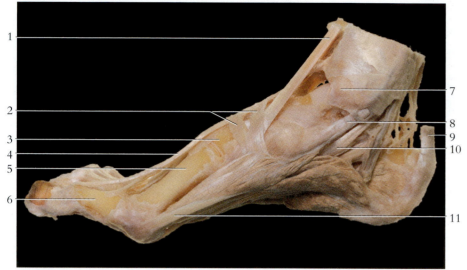

Figure 9.38 A medial view of the right foot
1. Tendon of tibialis anterior m.
2. Extensor retinaculum
3. Medial cuneiform
4. Tendon of extensor hallucis longus m.
5. First metatarsal bone
6. Proximal phalanx of hallux
7. Medial malleolus of tibia
8. Tendon of tibialis posterior m.
9. Tendo calcaneus
10. Tendon of flexor digitorum longus m.
11. Abductor hallucis m.

Figure 9.39 A lateral view of the right foot.
1. Tendon of tibialis anterior
2. Tendon of extensor digitorum longus m.
3. Tendon of fibularis tertius m.
4. Superior extensor retinaculum
5. Inferior extensor retinaculum
6. Lateral malleolus of fibula
7. Extensor digitorum brevis m.
8. Tendon of fibularis longus m.
9. Tendon of fibularis brevis m.
10. Calcaneus
11. Fifth metatarsal bone

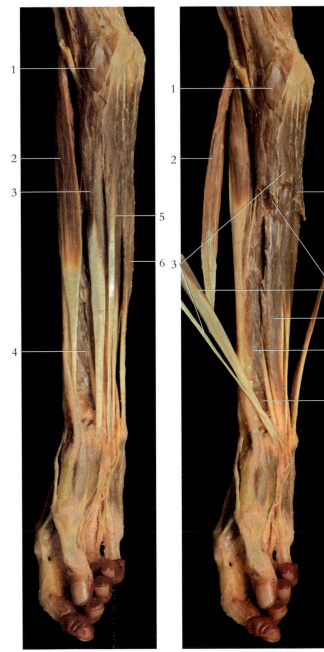

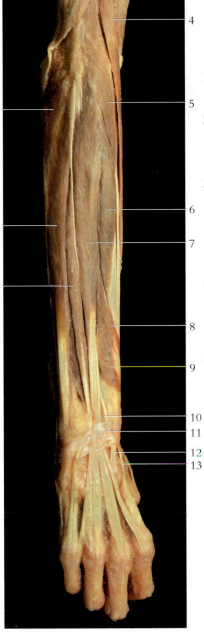

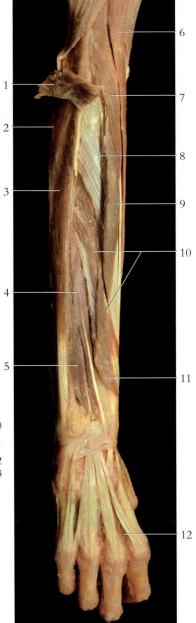

Figure 9.40 An anterior view of the superficial muscles of the right forearm.
1. Pronator teres m.
2. Brachioradialis m.
3. Flexor carpi radialis m.
4. Flexor pollicis longus m.
5. Palmaris longus m.
6. Flexor carpi ulnaris m.

Figure 9.41 An anterior view of the muscles of the right forearm.
1. Pronator teres m.
2. Brachioradialis m. (cut and reflected)
3. Palmaris longus m. (cut and reflected)
4. Flexor carpi ulnaris m. (cut)
5. Flexor carpi radialis m. (cut and reflected)
6. Flexor digitorum superficialis m.
7. Flexor pollicis longus m.
8. Pronator quadratus m.

Figure 9.42 A posterior view of the superficial muscles of the right forearm.
1. Anconeus m.
2. Extensor carpi ulnaris m.
3. Extensor digiti minimi m.
4. Brachioradialis m.
5. Extensor carpi radialis longus m.
6. Extensor carpi radialis brevis m.
7. Extensor digitorum m.
8. Abductor pollicis longus m.
9. Extensor pollicis brevis m.
10. Extensor pollicis longus m.
11. Extensor retinaculum
12. Tendon of extensor carpi radialis brevis
13. Tendon of extensor carpi radialis longus

Figure 9.43 A posterior view of the deep muscles of the right forearm.
1. Extensor digitorum m. (cut and reflected)
2. Anconeus m.
3. Extensor carpi ulnaris m.
4. Extensor pollicis longus m.
5. Extensor indicis m.
6. Brachioradialis m.
7. Extensor carpi radialis longus m.
8. Supinator m.
9. Extensor carpi radialis brevis m.
10. Abductor pollicis longus m.
11. Extensor pollicis brevis m.
12. Dorsal interosseous m.

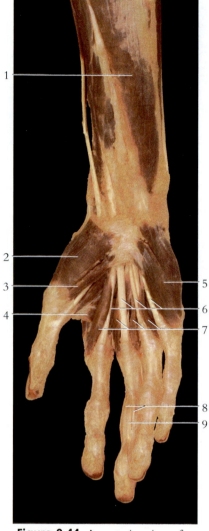

Figure 9.44 An anterior view of right hand
1. Flexor carpi ulnaris m.
2. Abductor pollicis brevis m.
3. Flexor pollicis brevis m.
4. Adductor pollicis m.
5. Hypothenar mm.
6. Tendons of flexor digitorum superficialis
7. Lumbrical mm.
8. Flexor digitorum superficialis tendon (bifurcated for insertion)
9. Tendon of flexor digitorum profundus

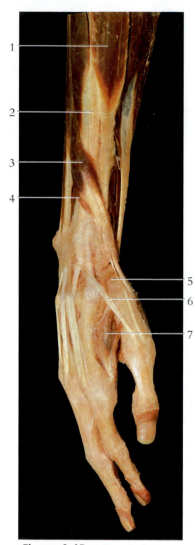

Figure 9.45 A lateral view of right hand
1. Extensor carpi radialis longus m.
2. Tendon of extensor carpi radialis brevis m.
3. Abductor pollicis longus m.
4. Extensor pollicis brevis m.
5. Anatomical snuff box
6. Tendon of extensor pollicis longus m.
7. First dorsal interosseus m.

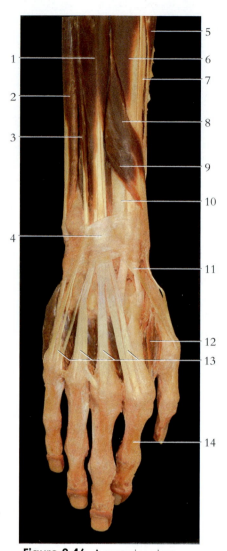

Figure 9.46 A posterior view of right hand
1. Extensor digitorum m.
2. Extensor carpi ulnaris m.
3. Extensor digiti minimi m.
4. Extensor retinaculum
5. Brachioradialis m.
6. Extensor carpi radialis brevis m.
7. Tendon of extensor carpi radialis longus m.
8. Abductor pollicis longus m.
9. Extensor pollicis brevis m.
10. Radius
11. Tendon of extensor pollicis longus m.
12. First dorsal interosseus m.
13. Extensor digitorum tendons
14. Extensor expansion

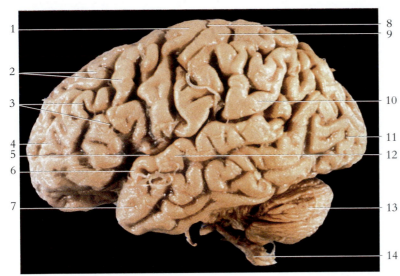

Figure 9.47 A lateral view of the brain.

1. Primary motor cerebral cortex
2. Gyri
3. Sulci
4. Frontal lobe of cerebrum
5. Lateral sulcus
6. Olfactory cerebral cortex
7. Temporal lobe of cerebrum
8. Central sulcus
9. Primary sensory cerebral cortex
10. Parietal lobe of cerebrum
11. Occipital lobe of cerebrum
12. Auditory cerebral cortex
13. Cerebellum
14. Medulla oblongata

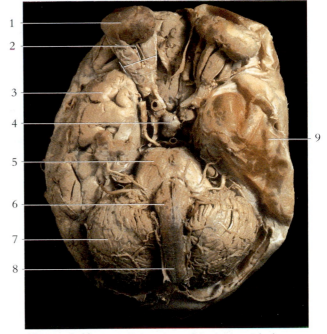

Figure 9.48 An inferior view of the brain with the eyes and part of the meninges still intact.

1. Eyeball
2. Muscles of the eye
3. Temporal lobe of cerebrum
4. Pituitary gland
5. Pons
6. Medulla oblongata
7. Cerebellum
8. Spinal cord
9. Dura mater

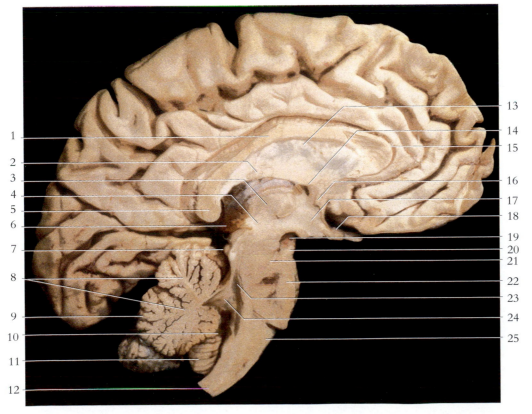

Figure 9.49 A sagittal view of the brain.

1. Body of corpus callosum
2. Crus of fornix
3. Third ventricle
4. Posterior commissure
5. Splenium of corpus callosum
6. Pineal body
7. Inferior colliculus
8. Arbor vitae of cerebellum
9. Vermis of cerebellum
10. Choroid plexus of fourth ventricle
11. Tonsilla of cerebellum
12. Medulla oblongata
13. Septum pellucidum
14. Intraventricular foramen
15. Genu of corpus callosum
16. Anterior commissure
17. Hypothalmus
18. Optic chiasma
19. Oculomotor nerve
20. Cerebral peduncle
21. Midbrain
22. Pons
23. Mesencephalic (cerebral) aqueduct
24. Fourth ventricle
25. Pyramid of medulla oblongata

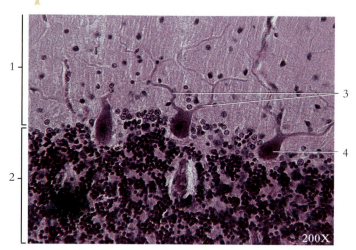

Figure 9.50 A photomicrograph of purkinje neurons from the cerebellum.
1. Molecular layer of cerebellar cortex
2. Granular layer of cerebellar cortex
3. Dendrites of Purkinje cell
4. Purkinje cell body

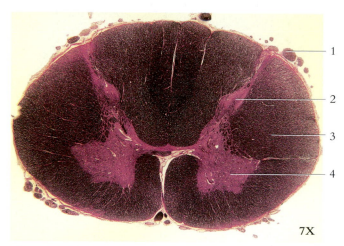

Figure 9.51 A transverse section of the spinal cord.
1. Posterior (dorsal) root of spinal nerve
2. Posterior (dorsal) horn (gray matter)
3. Spinal cord tract (white matter)
4. Anterior (ventral) horn (gray matter)

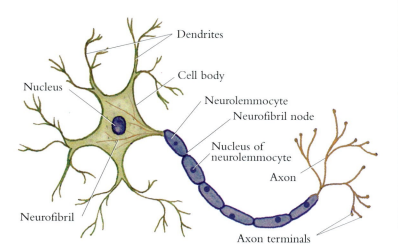

Dendrites

Cell body

Nucleus

Neurolemmocyte

Neurofibril node

Nucleus of neurolemmocyte

Axon

Neurofibril

Axon terminals

Figure 9.52 The structure of a myelinated neuron.

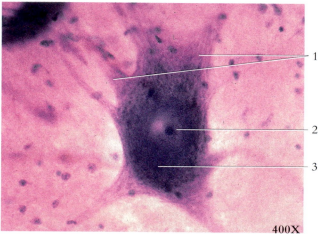

Figure 9.53 A photomicrograph of a neuron.
1. Cytoplasmic extensions   3. Cell body of neuron
2. Nucleolus

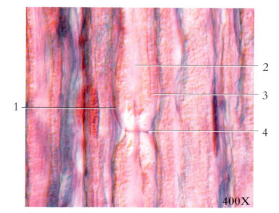

Figure 9.54 The histology of a myelinated nerve.

1. Endoneurium   4. Neurofibril node
2. Axon   (node of Ranvier)
3. Myelin layer

Figure 9.55 A transverse section of a nerve.
1. Perineurium   3. Bundle of axons
2. Epineurium

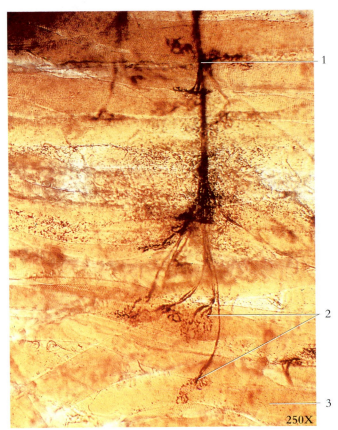

Figure 9.56 The neuromuscular junction.
1. Motor nerve   3. Skeletal muscle fiber
2. Motor end plates

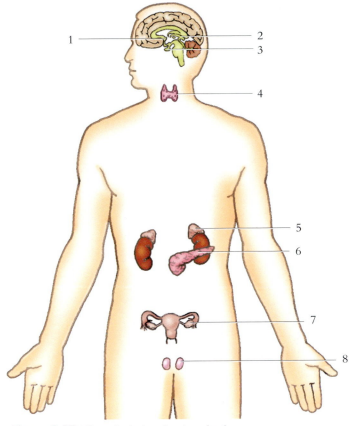

Figure 9.57 The principal endocrine glands.
1. Hypothalamus   5. Adrenal (suprarenal) gland
2. Pineal body   6. Pancreas
3. Pituitary gland   7. Ovary
4. Thyroid and parathyroid glands   8. Testis

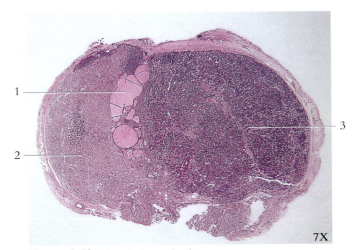

Figure 9.58 The pituitary gland.
1. Pars intermedia (adenohypophysis)
2. Pars nervosa (neurohypophysis)
3. Pars distalis (adenohypophysis)

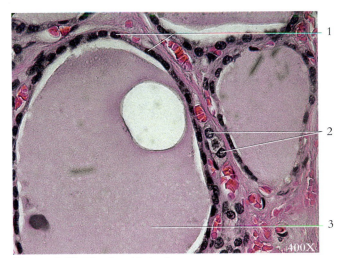

Figure 9.59 The thyroid gland.
1. Follicle cells
2. C cells
3. Colloid within follicle

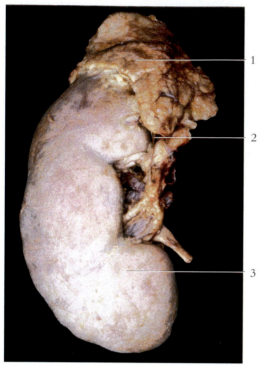

Figure 9.60 The adrenal (suprarenal) gland.
1. Adrenal gland
2. Inferior suprarenal artery
3. Kidney

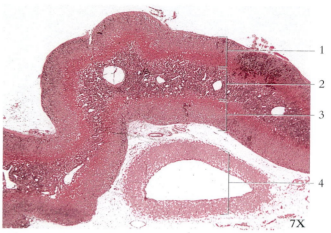

Figure 9.61 The adrenal gland.
1. Adrenal cortex
2. Adrenal medulla
3. Adrenal cortex
4. Blood vessel

7X

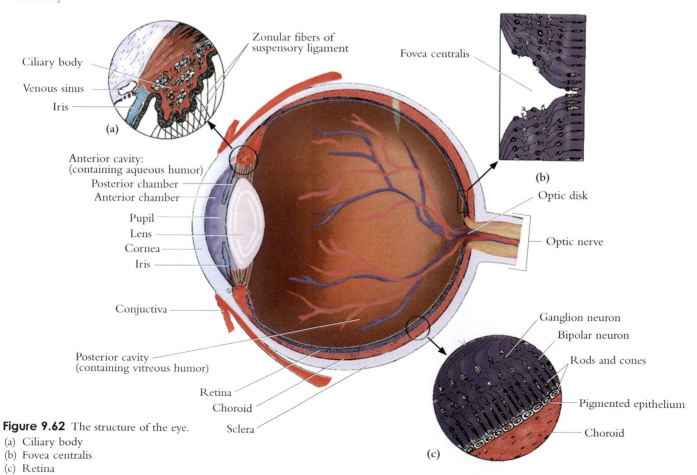

Zonular fibers of
suspensory ligament

Ciliary body

Venous sinus

Iris

Fovea centralis

(a)

Anterior cavity:
(containing aqueous humor)

Posterior chamber

Anterior chamber

Pupil

Lens

Cornea

Iris

Conjuctiva

Posterior cavity
(containing vitreous humor)

Retina

Choroid

Sclera

(b)

Optic disk

Optic nerve

Ganglion neuron

Bipolar neuron

Rods and cones

Pigmented epithelium

Choroid

(c)

Figure 9.62 The structure of the eye.
(a) Ciliary body
(b) Fovea centralis
(c) Retina

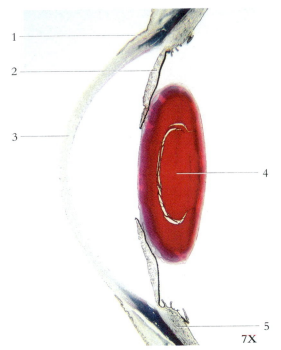

Figure 9.63 The anterior portion of the eye.
1. Conjunctivia   4. Lens
2. Iris   5. Ciliary body
3. Cornea

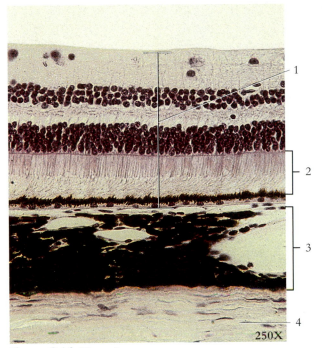

Figure 9.64 The retina.
1. Retina   3. Choroid
2. Rods and cones   4. Sclera

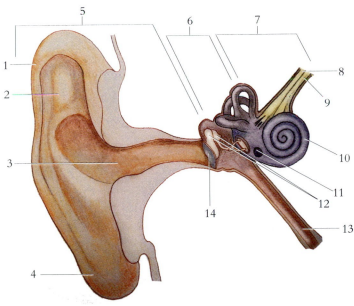

Figure 9.65 The structure of the ear.
1. Helix   8. Facial nerve
2. Auricle   9. Vestibulocochlear nerve
3. External auditory canal 10. Cochlea
4. Earlobe   11. Vestibular (oval) window
5. Outer ear   12. Auditory ossicles
6. Middle ear   13. Auditory tube
7. Inner ear   14. Tympanic membrane

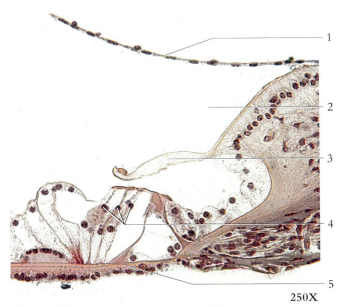

Figure 9.66 The spiral organ (organ of Corti).
1. Vestibular membrane
2. Cochlear duct
3. Tectorial membrane
4. Hair cells
5. Basilar membrane

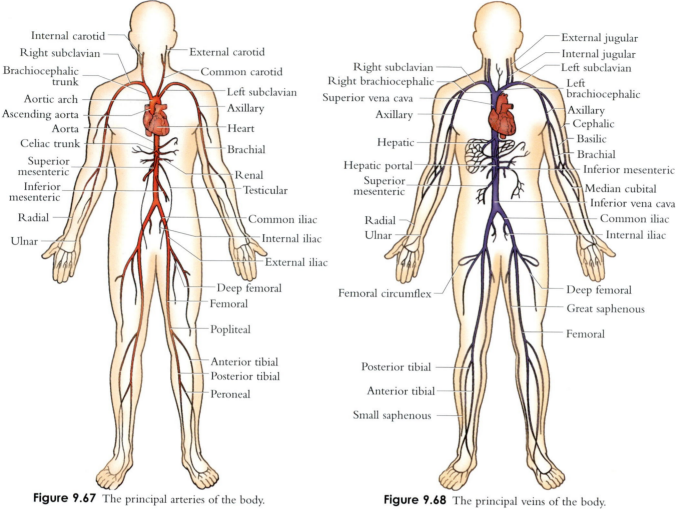

Figure 9.67 The principal arteries of the body.

Internal carotid
Right subclavian
Brachiocephalic trunk
Aortic arch
Ascending aorta
Aorta
Celiac trunk
Superior mesenteric
Inferior mesenteric
Radial
Ulnar

External carotid
Common carotid
Left subclavian
Axillary
Heart
Brachial
Renal
Testicular
Common iliac
Internal iliac
External iliac
Deep femoral
Femoral
Popliteal
Anterior tibial
Posterior tibial
Peroneal

Figure 9.68 The principal veins of the body.

Right subclavian
Right brachiocephalic
Superior vena cava
Axillary
Hepatic
Hepatic portal
Superior mesenteric
Radial
Ulnar

Femoral circumflex

Posterior tibial
Anterior tibial
Small saphenous

External jugular
Internal jugular
Left subclavian
Left brachiocephalic
Axillary
Cephalic
Basilic
Brachial
Inferior mesenteric
Median cubital
Inferior vena cava
Common iliac
Internal iliac
Deep femoral
Great saphenous
Femoral

Figure 9.69 The position of the heart within the pericardium.

1. Mediastinum
2. Right lung
3. Pericardium
4. Diaphragm
5. Liver
6. Left lung

Figure 9.70 Anterior view of the heart and associated structures

1. Right vagus nerve
2. Right brachiocephalic vein
3. Superior vena cava
4. Right phrenic nerve
5. Ascending aorta
6. Pericardium (cut)
7. Right ventricle of heart
8. Brachiocephalic artery
9. Left brachiocephalic vein
10. Aortic arch
11. Left phrenic nerve
12. Left ventricle of heart
13. Apex of heart

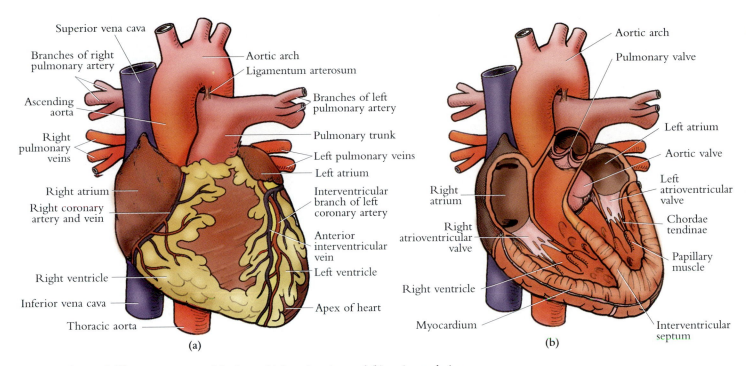

Superior vena cava
Branches of right pulmonary artery
Ascending aorta
Right pulmonary veins
Right atrium
Right coronary artery and vein
Right ventricle
Inferior vena cava
Thoracic aorta

Aortic arch
Ligamentum arterosum
Branches of left pulmonary artery
Pulmonary trunk
Left pulmonary veins
Left atrium
Interventricular branch of left coronary artery
Anterior interventricular vein
Left ventricle
Apex of heart

(a)

Aortic arch
Pulmonary valve
Left atrium
Aortic valve
Left atrioventricular valve
Chordae tendinae
Papillary muscle
Interventricular septum

Right atrium
Right atrioventricular valve
Right ventricle
Myocardium

(b)

Figure 9.71 The structure of the heart. (a) Anterior view and (b) an internal view.

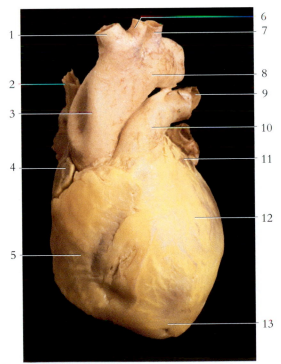

Figure 9.72 An anterior view of the heart and great vessels.

1. Brachiocephalic trunk
2. Superior vena cava
3. Ascending aorta
4. Right atrium
5. Right ventricle
6. Left common carotid artery
7. Left subclavian artery
8. Aortic arch
9. Pulmonary artery
10. Pulmonary trunk
11. Left atrium
12. Left ventricle
13. Apex of heart

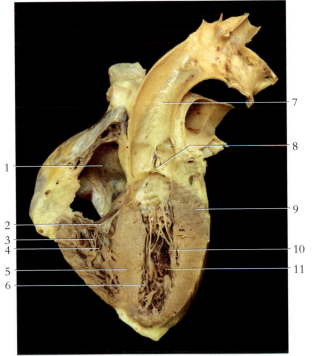

Figure 9.73 The internal structure of the heart.

1. Right atrium
2. Right atrioventricular valve
3. Chordae tendinae
4. Left ventricle
5. Interventricular septum
6. Trabeculae carneae
7. Ascending aorta
8. Aortic valve
9. Myocardium
10. Papillary muscle
11. Left ventricle

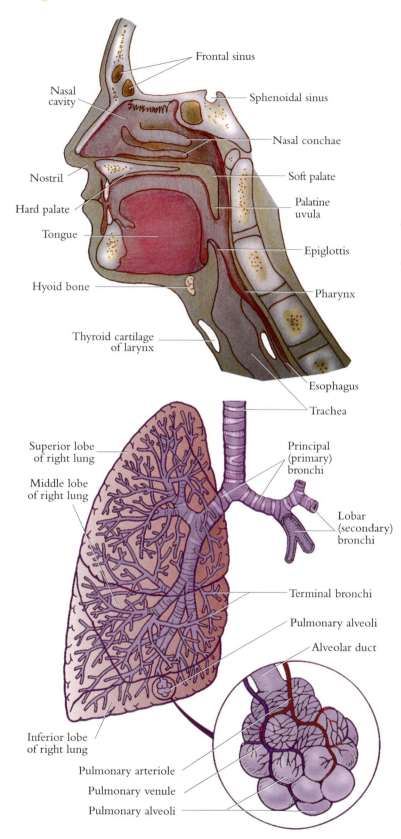

Figure 9.74 The structure of the respiratory system.

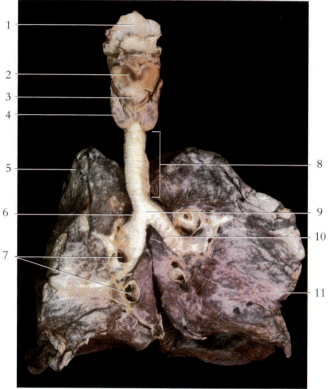

Figure 9.75 An anterior view of the larynx, trachea, and lungs.

1. Epiglottis
2. Thyroid cartilage
3. Cricoid cartilage
4. Thyroid gland
5. Right lung
6. Right principal (primary) bronchus
7. Pulmonary vessels
8. Trachea
9. Carina
10. Left principal (primary) bronchus
11. Left lung

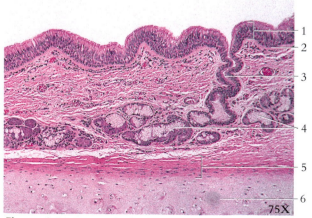

Figure 9.76 The tracheal wall.

1. Respiratory epithelium
2. Basement membrane
3. Duct of seromucous gland
4. Seromucous glands
5. Perichondrium
6. Hyaline cartilage

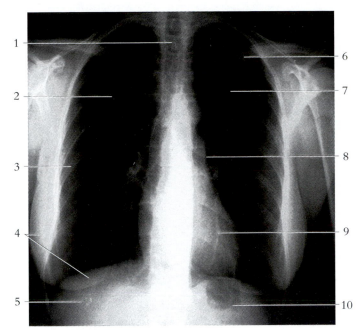

Figure 9.77 A radiograph of the thorax.
1. Thoracic vertebra   6. Clavicle
2. Right lung   7. Left lung
3. Rib   8. Mediastinum
4. Image of right breast   9. Heart
5. Diaphragm/liver   10. Diaphragm/stomach

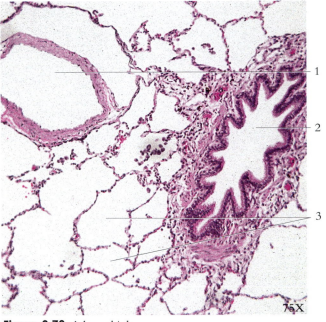

Figure 9.78 A bronchiole.
1. Pulmonary arteriole
2. Bronchiole
3. Pulmonary alveoli

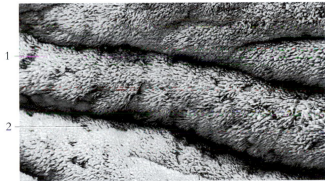

Figure 9.79 An electron micrograph of the lining of the trachea.
1. Cilia   2. Goblet cell

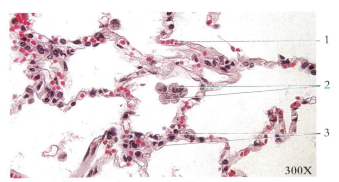

Figure 9.80 The pulmonary alveoli.
1. Capillary in alveolar wall   3. Type II pneumocytes
2. Macrophages

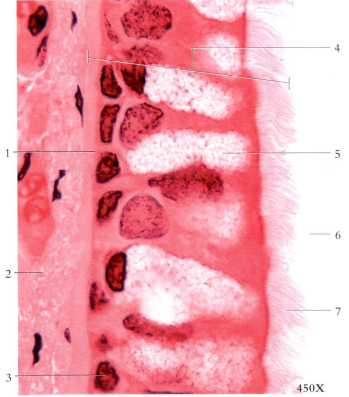

Figure 9.81 The bronchus.
1. Basement membrane   5. Goblet cell
2. Lamina propria   6. Lumen of bronchus
3. Nucleus   7. Cilia
4. Pseudostratified squamous epithelium

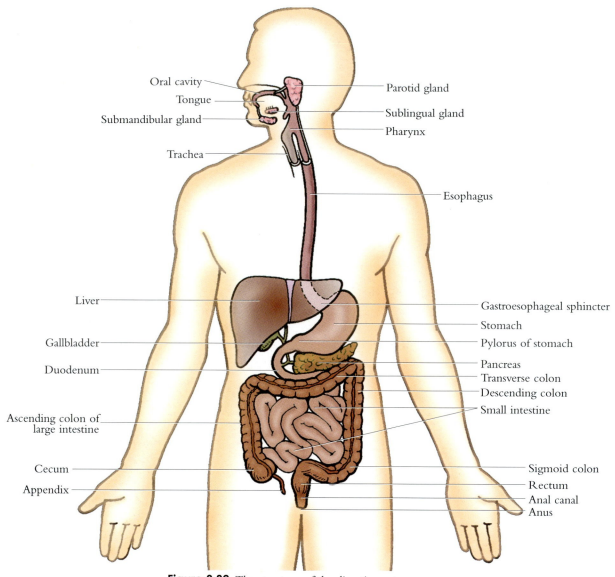

Figure 9.82 The structure of the digestive system.

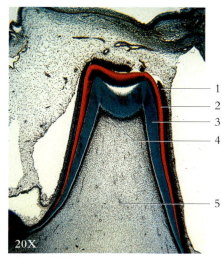

Figure 9.83 A developing tooth.

1. Ameloblasts   4. Odontoblasts
2. Enamel   5. Pulp
3. Dentin

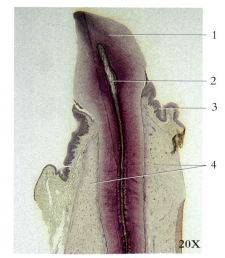

Figure 9.84 A mature tooth.

1. Dentin (enamel has   3. Gingiva
   been dissolved away)   4. Alveolar bone
2. Pulp

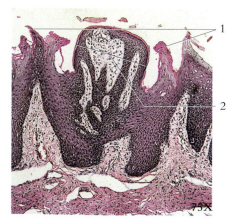

Figure 9.85 The filiform and fungiform papillae.

1. Filiform papillae
2. Fungiform papilla

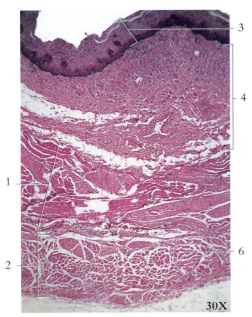

Figure 9.86 The wall of the esophagus.
1. Inner circular layer
   (muscularis externa)
2. Outer longitudinal layer
   (muscularis externa)
3. Mucosa
4. Submucosa
5. Muscularis externa

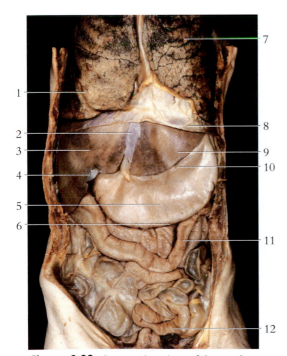

Figure 9.88 An anterior view of the trunk.

1. Right lung	7. Left lung
2. Falciform ligament	8. Diaphragm
3. Right lobe of liver	9. Left lobe of liver
4. Gallbladder	10. Lesser curvature of stomach
5. Body of stomach	11. Transverse colon
6. Greater curvature of stomach	12. Small intestine

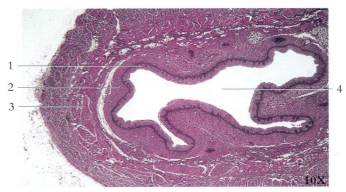

Figure 9.87 A transverse section of esophagus.
1. Mucosa   3. Muscularis
2. Submucosa   4. Lumen

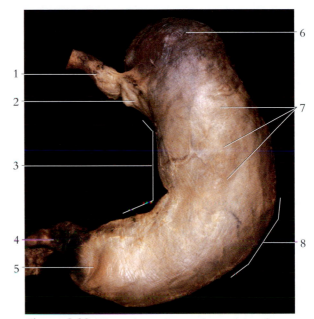

Figure 9.89 The major regions and structures of
the stomach.

1. Esophagus	5. Pylorus of stomach
2. Cardiac portion of stomach	6. Fundus of stomach
3. Lesser curvature of stomach	7. Body of stomach
4. Duodenum	8. Greater curvature of stomach

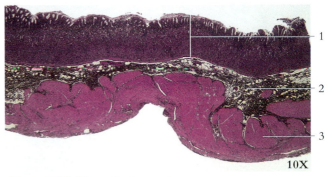

Figure 9.90 The wall of stomach.
1. Mucosa   3. Muscularis externa
2. Submucosa

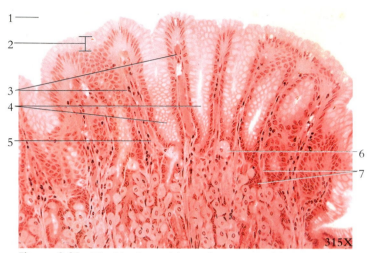

315X

Figure 9.91 The histology of the cardiac region of the stomach.
1. Lumen of stomach
2. Surface epithelium
3. Mucosal ridges
4. Gastric pits
5. Lamina propria
6. Parietal cells
7. Chief (zymogenic) cells

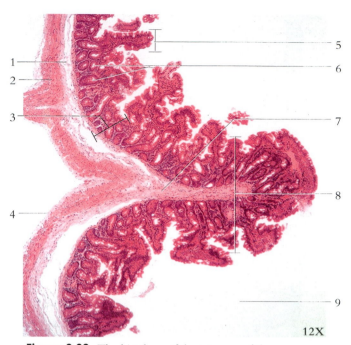

12X

Figure 9.92 The histology of the jejunum of the small intestine.
1. Submucosa
2. Circular and longitudinal muscles
3. Mucosa
4. Serosa
5. Villus
6. Intestinal glands
7. Submucosa
8. Plica circulares
9. Lumen of small intestine

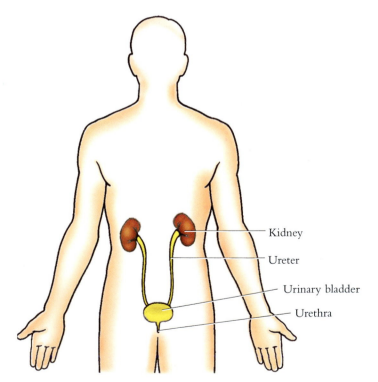

Kidney

Ureter

Urinary bladder

Urethra

Figure 9.93 The organs of the urinary system.

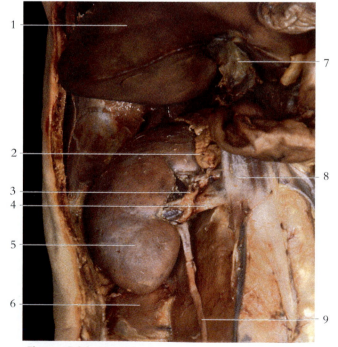

Figure 9.94 The kidney and ureter with overlying viscera removed.
1. Liver
2. Adrenal gland
3. Renal artery
4. Renal vein
5. Right kidney
6. Quadratus lumborum muscle
7. Gallbladder
8. Inferior vena cava
9. Ureter

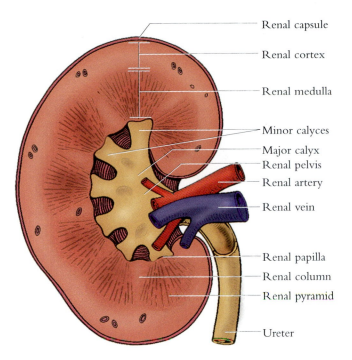

Renal capsule

Renal cortex

Renal medulla

Minor calyces

Major calyx

Renal pelvis

Renal artery

Renal vein

Renal papilla

Renal column

Renal pyramid

Ureter

Figure 9.95 The structure of the kidney.

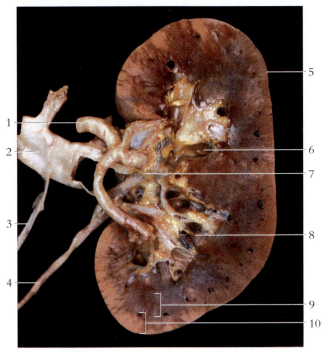

1

2

3

4

5

6

7

8

9

10

Figure 9.96 A coronal section of the left kidney.

1. Renal artery   6. Major calyx
2. Renal vein   7. Renal pelvis
3. Left testicular vein 8. Renal papilla
4. Ureter   9. Renal medulla
5. Renal capsule   10. Renal cortex

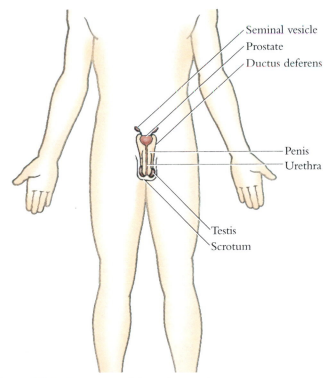

Seminal vesicle
Prostate
Ductus deferens

Penis
Urethra

Testis
Scrotum

Figure 9.97 The organs of the male reproductive system.

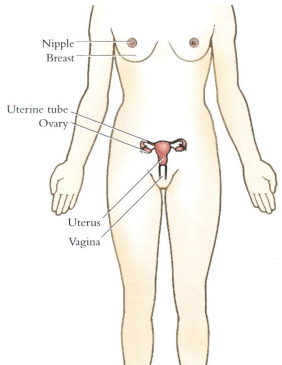

Nipple
Breast

Uterine tube
Ovary

Uterus

Vagina

Figure 9.98 The organs of the female reproductive system.

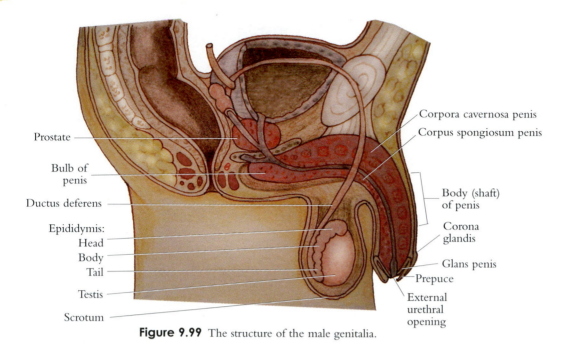

Prostate

Bulb of penis

Ductus deferens

Epididymis:

Head

Body

Tail

Testis

Scrotum

Corpora cavernosa penis

Corpus spongiosum penis

Body (shaft) of penis

Corona glandis

Glans penis

Prepuce

External urethral opening

Figure 9.99 The structure of the male genitalia.

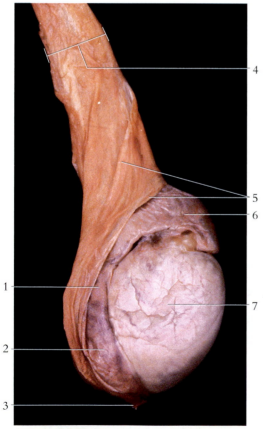

Figure 9.100 A testis and associated structures.
1. Body of epididymis
2. Tail of epididymis
3. Gubernaculum
4. Spermatic cord
5. Spermatic fascia
6. Head of epididymis
7. Testis

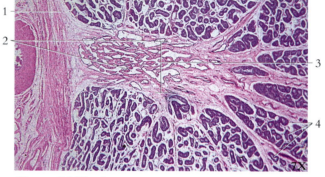

Figure 9.101 The testis.
1. Tunic albuginea
2. Tubules of rete testis
3. Mediastinum
4. Seminiferous tubules

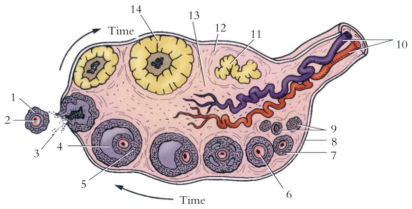

Time

Time

Figure 9.102 The structure of the ovary.
1. Corona radiata
2. Secondary oocyte
3. Ovulation
4. Follicular fluid within antrum
5. Cumulus oophorus
6. Oocyte
7. Follicular cells
8. Germinal epithelium
9. Primary follicles
10. Ovarian vessels
11. Corpus albicans
12. Ovarian cortex
13. Ovarian medulla
14. Corpus luteum

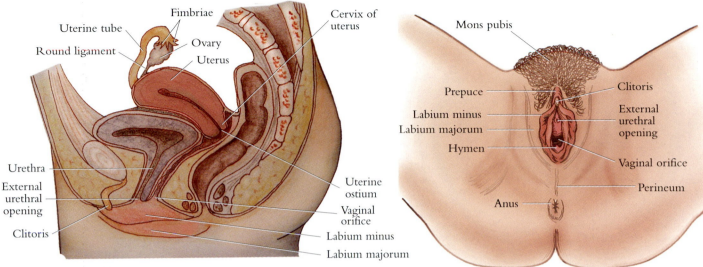

Figure 9.103 The external genitalia and internal reproductive organs of the female reproductive system.

Figure 9.104 The female external genitalia.

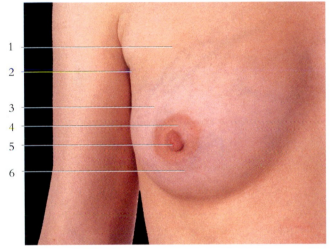

Figure 9.105 The surface anatomy of the female breast.
1. Pectoralis major muscle
2. Axilla
3. Lateral process of breast
4. Areola
5. Nipple
6. Breast (containing mammary glands)

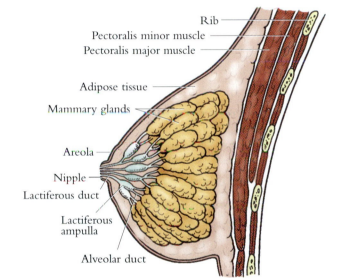

Figure 9.106 The mammary gland.

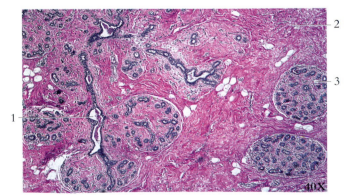

Figure 9.107 Mammary glands (non-lactating glands).
1. Interlobular duct
2. Interlobular connective tissue
3. Lobule of glandular tissue

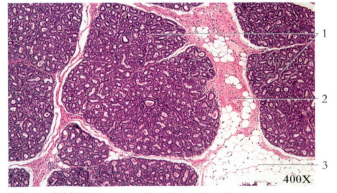

Figure 9.108 Mammary glands (lactating glands).
1. Lobules of glandular tissue
2. Intralobular connective tissue
3. Adipose cells

Glossary of Terms

abdomen – the portion of the trunk of the mammalian body located between the diaphragm and the pelvis, that contains the abdominal cavity and its visceral organs; one of the three principal body regions (head, thorax, and abdomen) of many animals.

abduction – a movement away from the axis or midline of the body; opposite of adduction, a movement of a digit away from the axis of a limb.

abiotic – without living organisms; nonliving portions of the environment.

abscission – the shedding of leaves, flowers, fruits, or other plant parts, usually following the formation of an abscission zone.

absorption – movement of a substance into a cell or an organism, or through a surface within an organism.

acapnia – a decrease in normal amount of CO_2 in the blood.

accommodation – a change in the shape of the lens of the eye so that vision is more acute; focusing for various distances.

acetone – an organic compound that may be present in the urine of diabetics; also called *ketone body.*

acetylcholine – a neurotransmitter chemical secreted at the terminal ends of many neurons, responsible for postsynaptic transmission; also called *ACh.*

acetylcholinesterase – an enzyme that breaks down acetylcholine; also called *AChE.*

Achilles tendon – see *tendo calcaneous.*

acid – a substance that releases hydrogen ions (H^+) in a solution.

acidosis – a disorder of body chemistry in which the alkaline substances of the blood are reduced in an amount below normal.

acoelomate – without a coelomic cavity; as in flatworms.

acoustic – referring to sound or the sense of hearing.

actin – a protein in muscle fibers that together with myosin is responsible for contraction.

action potential – the change in ionic charge propagated along the membrane of a neuron; the *nerve impulse.*

active transport – movement of a substance into or out of a cell from a lesser to a greater concentration, requiring a carrier molecule and expenditure of energy.

adaptation – structural, physiological, or behavioral traits of an organism that promote its survival and contributes to its ability to reproduce under environmental conditions.

adduction – a movement toward the axis or midline of the body; opposite of abduction, a movement of a digit toward the axis of a limb.

adenohypophysis – anterior pituitary.

adenoid – paired lymphoid structures in the nasopharynx; also called *pharyngeal tonsils.*

adenosine triphosphate (ATP) – a chemical compound that provides energy for cellular use.

adhesion – an attraction between unlike substances.

adipose – fat, or fat-containing, such as adipose tissue.

adrenal glands – endocrine glands; one superior to each kidney; also called *suprarenal glands.*

adventitious root – Supportive root developing from the stem of a plant.

aerobic – requiring free O_2 for growth and metabolism as in the case of certain bacteria also called *aerobes.*

agglutination – clumping of cells; particular reference to red blood cells in an antigen-antibody reaction.

aggregate fruit – ripened ovaries from a single flower with several separate carpels.

aggression – provoking, domineering behavior.

alga (pl. algae) – any of a diverse group of aquatic photosynthesizing organisms that are either unicellular or multicellular; algae comprise the phytoplankton and seaweeds of the Earth.

alkaline – a substance having a pH greater than 7.0; *basic.*

allantois – an extraembryonic membranous sac that forms blood cells and gives rise to the fetal umbilical arteries and vein. It also contributes to the formation of the urinary bladder.

allele – an alternative form of a gene occurring at a given chromosome site, or *locus.*

all-or-none response – functioning completely when exposed to a stimulus of threshold strength; applies to action potentials through neurons and muscle fiber contraction.

alpha helix – right-handed spiral typical in proteins and DNA.

alternation of generations – two-phased life cycle characteristic of many plants in which there are sporophyte and gametophyte generations.

altruism – behavior benefiting other organisms without regard to its possible advantage or detrimental effect on the performer.

alveolus – An individual air capsule within the lung. Pulmonary alveoli are the basic functional units of respiration. Also, the socket that secures a tooth (dental alveolus).

amino acid – a unit of protein that contains an amino group (NH_2) and an acid group (COOH).

amnion – a membrane that surrounds the fetus to contain the amniotic fluid.

amniote – an animal that has an amnion during embryonic development; reptiles, birds, and mammals.

amoeba – protozoans that move by means of pseudopodia.

amphiarthrosis – a slightly moveable joint in a functional classification of joints.

anaerobic respiration – metabolizing and growing in the absence of oxygen.

analogous – similar in function regardless of developmental origin; generally in reference to similar adaptations.

anatomical position – the position in human anatomy in which there is an erect body stance with the eyes directed forward, the arms at the sides, and the palms of the hands facing forward.

anatomy – the branch of science concerned with the structure of the body and the relationship of its organs.

angiosperm – flowering plant, having double fertilization resulting in development of specialized seeds within fruits.

annual – a flowering plant that completes its entire life cycle in a single year or growing season.

annual ring – yearly growth demarcation in woody plants formed by build-up of secondary xylem.

annulus – a ring-like segment, such as body rings on an earthworm.

antebrachium – the forearm.

antenna – a sensory appendage on many species of invertebrate animals.

anterior (ventral) – toward the front; the opposite of *posterior (dorsal).*

anther – the position of a plant stamen in which pollen is produced.

antheridium – male reproductive organ in certain nonseed plants and algae where motile sperm are produced.

anticodon – three ("a triplet") nucleotides sequence in transfer RNA that pairs with a complementary codon (triplet) in messenger RNA.

antigen – a foreign material, usually a protein, that triggers the immune system to produce antibodies.

anus – the terminal end of the gastrointestinal tract, opening of the anal canal.

aorta – the major systemic vessel of the arterial portion of the circulatory system, emerging from the left ventricle.

apical meristem – embryonic plant tissue in the tip of a root, bud, or shoot where continual cell divisions cause growth in length.

apocrine gland – a type of sweat gland that functions in evaporative cooling.

apopyle – opening of the radial canal into the spongocoel of sponges.

appeasement – submission behavior, usually soliciting an end of aggression.

appendix – a short pouch that attaches to the cecum.

aqueous humor – the watery fluid that fills the anterior and posterior chambers of the eye.

arachnoid – the web-like middle covering (meninx) of the central nervous system.

arbor vitae – the branching arrangement of white matter within the cerebellum.

archaebacteria – organisms within the kingdom Monera that represents an early group of simple life forms.

archegonium – female reproductive organ in certain nonseed plants; a gametangium where eggs are produced.

archenteron – the principal cavity of an embryo during the gastrula stage. Lined with endoderm, the archenteron develops into the digestive tract.

areola – the pigmented ring around the nipple.

arteries – a blood vessels that carry blood away from the heart.

articular cartilage – a hyaline cartilaginous covering over the articulating surface of bones of synovial joints.

ascending colon – the portion of the large intestine between the cecum and the hepatic (right colic) flexure.

asexual – lacking distinct sexual organs and lacking the ability to produce gametes.

aster – minute rays of microtubules at the ends of the spindle apparatus in animal cells during cell division.

asymmetry – not symmetrical.

atom – the smallest unit of an element that can exist and still have the properties of the element; collectively, atoms form molecules in a compound.

atomic number – the weight of the atom of a particular element.

atomic weight – the number of protons together with the number of neutrons within the nucleus of an atom.

ATP (adenosine triphosphate) – a compound of adenine, ribose, and three phosphates; it is the energy source for most cellular processes.

atrium – either of two superior chambers of the heart that receive venous blood.

atrophy – a wasting away or decrease in size of a cell or organ.

auditory tube – a narrow canal that connects the middle ear chamber to the pharynx; also called the *eustachian canal.*

autonomic – self-governing; pertaining to the division of the nervous system that controls involuntary activities.

autosome – a chromosome other than a sex chromosome.

autotroph – an organism capable of synthesizing its own organic molecules (food) from inorganic molecules.

axilla – the depressed hollow under the arm; the armpit.

axillary bud – a group of meristematic cells at the junction of a leaf and stem that develops branches or flowers; also called *lateral bud*.

axon – The elongated process of a neuron (nerve cell) that transmits an impulse away from the cell body.

bacillus (pl. bacilli) – a rod-shaped bacterium.

bacteria – prokaryotes within the kingdom Monera, lacking the organelles of eukaryotic cells.

bark – outer tissue layers of a tree consisting of cork, cork cambium, cortex, and phloem.

basal – at or near the base or point of attachment, as of a plant shoot.

base – a substance that contributes or liberates hydroxide ions in a solution.

basement membrane – a thin sheet of extracellular substance to which the basal surfaces of membranous epithelial cells are attached.

basidia – club-shaped reproductive structures of club fungi that produce basidiospores during sexual reproduction.

basophil – a granular leukocyte that readily stains with basophilic dye.

belly – the thickest circumference of a skeletal muscle.

benign – nonmalignant; a confined tumor.

berry – a simple fleshy fruit.

biennial – a plant that lives through two growing seasons; generally, these plants have vegetative growth during the first season, and flower and set seed during the second season.

bilateral symmetry – the morphologic condition of having similar right and left halves.

binary fission – a process of sexual reproduction that does not involve a mitotic spindle.

binomial nomenclature – assignment of two names to an organism, the first of which is the genus and the second the specific epithet; together constituting the scientific name.

biome – a major climax community characterized by a particular group of plants and animals.

biosphere – the portion of the earth's atmosphere and surface where living organisms exist.

biotic – pertaining to aspects of life, especially to characteristics of ecosystems.

bisexual flower – a flower that contains both male and female sexual structures.

blade – the broad expanded portion of a leaf.

blastocoel – the cavity of any blastula.

blastula – stage of embryonic development of animals near the end of cleavage but before gastrulation.

blood – the fluid connective tissue that circulates through the cardiovascular system to transport substances throughout the body.

bolus – a moistened mass of food that is swallowed from the oral cavity into the pharynx.

bone – an organ composed of solid, rigid connective tissue, forming a component of the skeletal system.

Bowman's capsule – see *glomerular capsule*.

brain – the enlarged superior portion of the central nervous system, located in the cranial cavity of the skull.

brainstem – the portion of the brain consisting of the medulla oblongata, pons, and midbrain.

bronchial tree – the bronchi and their branching bronchioles.

bronchiole – a small division of a bronchus within the lung.

bronchus – a branch of the trachea that leads to a lung.

buccal cavity – the mouth, or oral cavity.

budding – a type of asexual reproduction in which outgrowths from the parent plant pinch off to live independently or may remain attached to form colonies.

buffer – a compound or substance that prevents large changes in the pH of a solution.

bulb – a thickened underground stem often enclosed by enlarged, fleshy leaves containing stored food.

bursa – a sac-like structure filled with synovial fluid, which occurs around joints.

buttock – the rump or fleshy mass on the posterior aspect of the lower trunk, formed primarily by the gluteal muscles.

calorie – the heat required to raise one kilogram of water one degree centigrade.

calyx – a cup-shaped portion of the renal pelvis that encircles renal papillae; the collective term for the sepals of a flower.

cambium – the layer of meristematic tissue in roots and stems of many vascular plants that continues to produce tissue.

cancellous bone – spongy bone; bone tissue with a lattice-like structure.

capillary – a microscopic blood vessel that connects an arteriole and a venule; the functional unit of the circulatory system.

carapace – protective covering over the dorsal part of the body of certain crustaceans and turtles.

carcinogenic – stimulating or causing the growth of a malignant tumor, or cancer.

carnivore – any animal that feeds upon another; especially, flesh-eating mammal.

carpus – the proximal portion of the hand that contains the carpal bones.

carrying capacity – the maximum number of organisms of a species that can be maintained indefinitely in an ecosystem.

cartilage – a type of connective tissue with a solid elastic matrix.

catalyst – a chemical, such as an enzyme, that accelerates the rate of a reaction of a chemical process but is not used up in the process.

caudal – referring to a position more toward the tail.

cecum – the pouch-like portion of the large intestine to which the ileum of the small intestine is attached.

cell – the structural and functional unit of an organism; the smallest structure capable of performing all the functions necessary for life.

cell wall – a rigid protective structure of a plant cell surrounding the cell (plasma) membrane; often composed of cellulose fibers embedded in a polysaccharide/protein matrix.

cellular respiration – the reactions of glycolysis, Krebs cycle, and electron transport system that provide cellular energy and accompanying reactions to produce ATP.

cellulose – a polysaccharide produced as fibers that forms a major part of the rigid cell wall around a plant cell.

central nervous system (CNS) – the brain and the spinal cord.

centromere – a portion of the chromosome to which a spindle fiber attaches during mitosis or meiosis.

centrosome – a dense body near the nucleus of a cell that contains a pair of centrioles.

cephalothorax – fusion of the head and thoracic regions, characteristic of certain arthropods.

cercaria – larva of trematodes (flukes).

cerebellum – the portion of the brain concerned with the coordination of movements and equilibrium.

cerebrospinal fluid (CSF) – a liquid that buoys and cushions the central nervous system.

cerebrum – the largest portion of the brain, composed of the right and left hemispheres.

cervical – pertaining to the neck or a neck-like portion of an organ.

chelipeds – front pair of pincer-like legs in most decapod crustaceans, adapted for seizing and crushing.

chitin – strong, flexible polysaccharide forming the exoskeleton of arthropods.

chlorophyll – green pigment in photosynthesizing organisms that absorbs energy from the sun.

chloroplast – a membrane-enclosed organelle that contains chlorophyll and is the site of photosynthesis.

choanae – the two posterior openings from the nasal cavity into the nasopharynx.

cholesterol – a lipid used in the synthesis of steroid hormones.

chondrocyte – a cartilage cell.

chorion – an extraembryonic membrane that participates in the formation of the placenta.

choroid – the vascular, pigmented middle layer of the wall of the eye.

chromatin – thread-like network of DNA and proteins within the nucleus.

chromosome – structure in the nucleus that contains the genes for genetic expression.

chyme – the mass of partially digested food that passes from the stomach into the duodenum of the small intestine.

cilia – microscopic, hair-like processes that move in a wave-like manner on the exposed surfaces of certain epithelial cells.

ciliary body – a portion of the choroid layer of the eye that secretes aqueous humor and contains the ciliary muscle.

ciliates – protozoans that move by means of cilia.

circadian rhythm – a daily physiological or behavioral event, occurring on an approximate 24-hour cycle.

circumduction – a cone-like movement of a body part, such that the distal end moves in a circle while the proximal portion remains relatively stable.

clitoris – a small, erectile structure in the vulva of the female.

cochlea – the spiral portion of the inner ear that contains the spiral organ (organ of corti).

climax community – the final stable stage in succession.

clone – asexually produced organisms having a consistent genetic constitution.

cnidarian – small aquatic organisms having radial symmetry and stinging cells with nematocysts.

cocoon – protective, or resting, stage of development in certain invertebrate animals.

codon – a "triplet" of three nucleotides in RNA that directs the placement of an amino acid into a polypeptide chain.

coelom – body cavity of higher animals, containing visceral organs.

collar cells – flagella-supporting cells in the inner layer of the wall of sponges.

colon – the first portion of the large intestine.

common bile duct – a tube that is formed by the union of the hepatic duct and cystic duct; transports bile to the duodenum.

compact bone – tightly packed bone that is superficial to spongy bone; also called *dense bone*.

competition – interaction between individuals of the same or different species for a mutually necessary resource.

complete flower – a flower that has four whorls of floral components including sepals, petals, stamens, and carpels.

compound eye – arthropod eye consisting of multiple lenses.

compound leaf – a leaf blade divided into distinct leaflets.

condyle – a rounded process at the end of a long bone that forms an articulation.

conidia – spores produced by fungi during asexual reproduction.

conifer – a cone-bearing woody seed plant, such as pine, fir, and spruce

conjugation – sexual union in which the nuclear material of one cell enters another.

connective tissue – one of the four basic tissue types within an animal's body. It is a binding and supportive tissue with abundant matrix.

consumer – an organism that derives nutrients by feeding upon another.

control – a sample in an experiment that undergoes all the steps in the experiment except the one being investigated.

convergent evolution – the evolution of similar structures in different groups of organisms exposed to similar environments.

coral – a cnidarian that has a calcium carbonate skeleton whose remains contribute to form reefs.

cork – the protective outer layer of bark of many trees, composed of dead cells that may be sloughed off.

cornea – the transparent convex, anterior portion of the outer layer of the vertebrate eye.

cortex – the outer layer of an organ such as the convoluted cerebrum, adrenal gland, or kidney.

costal cartilage – the cartilage that connects the ribs to the sternum.

cranial – pertaining to the cranium.

cranial nerve – one of 12 pairs of nerves that arise from the inferior surface of the brain.

cranium – the bones of the skull that enclose the brain and support the organs of sight, hearing, and balance.

crossing over – the exchange of corresponding chromatid segments of genetic material of homologous chromosomes during synapsid of meiosis I.

cuticle – wax-like covering on the epidermis of nonwoody plants to prevent water loss.

cyanobacteria – photosynthetic prokaryotes that have chlorophyll and release oxygen.

cytokinesis – division of the cellular cytoplasm.

cytology – the science dealing with the study of cells.

cytoplasm – the protoplasm of a cell located outside of the nucleus.

cytoskeleton – protein filaments throughout the cytoplasm of certain cells that help maintain the cell shape.

deciduous – plants that seasonally shed their leaves.

dendrite – a nerve cell process that transmits impulses toward a neuron cell body.

denitrifying bacteria – single-cellular organisms that convert nitrate to atmospheric nitrogen.

dentin – the principal substance of a tooth, covered by enamel over the crown and by cementum on the root.

dermis – the second, or deep, layer of skin beneath the epidermis.

descending colon – the segment of the large intestine that descends on the left side from the level of the spleen to the level of the left iliac crest.

detritus – non-living organic matter important in the nutrient cycle in soil formation.

diaphragm – a flat dome of skeletal muscle and connective tissue that separates the thoracic and abdominal cavities, in mammals.

diaphysis – the body, or shaft, of a long bone.

diarthrosis – a freely movable joint in a functional classification of joints.

diastole – the portion of the cardiac cycle during which the ventricular heart chamber wall is relaxed.

diatoms – aquatic unicellular algae characterized by a cell wall composed of two silica-impregnated valves.

dicot – a kind of angiosperm characterized by the presence of two cotyledons in the seed; also called *dicotyledon*.

diffusion – movement of molecules from an area of greater concentration to an area of lesser concentration.

dihybrid cross – a breeding experiment in which parental varieties differing in two traits are mated.

dimorphism – two distinct forms within a species, with regard to size, color, organ structure, and so on.

diphyodent – two sets of teeth, deciduous and permanent.

diploid – having two copies of each different chromosome, pairs of homologous chromosomes (2N).

distal – away from the midline or origin; the opposite of *proximal*.

dominant – a hereditary characteristic that expresses itself even when the genotype is heterozygous.

dormancy – a period of suspended activity and growth.

dorsal – pertaining to the back or posterior portion of a body part; the opposite of *ventral*.

double helix – a double spiral used to describe the three-dimensional shape of DNA.

ductus deferens – a tube that carries spermatozoa from the epididymis to the ejaculatory duct; also called the *vas deferens* or *seminal duct*.

duodenum – the first portion of the small intestine.

dura mater – the outermost meninx covering the central nervous system.

eccrine gland – a sweat gland that functions in body cooling.

ecology – the study of the relationship of organisms and the physical environment and their interactions.

ecosystem – a biological community and its associated abiotic environment.

ectoderm – the outermost of the three primary embryonic germ layers.

edema – an excessive retention of fluid in the body tissues.

effector – an organ such as a gland or muscle that responds to motor stimulation.

efferent – conveying away from the center of an organ or structure.

ejaculation – the discharge of semen from the male urethra during climax.

electrocardiogram – a recording of the electrical activity that accompanies the cardiac cycle; also called *ECG* or *EKG*.

electroencephalogram – a recording of the brain wave pattern; also called *EEG*.

electromyogram – a recording of the activity of a muscle during contraction; also called *EMG*.

electrolyte – a solution that conducts electricity by means of charged ions.

electron – the unit of negative electricity.

element – a structure composed of only one type of atom (e.g., carbon, hydrogen, oxygen).

embryo – a plant or an animal at an early stage of development.

emulsification – the process of dispersing one liquid in another.

enamel – the outer, dense substance covering the crown of a tooth.

endocardium – the fibrous lining of the heart chambers and valves.

endochondral bone – bones that form as hyaline cartilage models first and then are ossified.

endocrine gland – a hormone-producing gland that secretes directly into the blood or body fluids.

endoderm – the innermost of the three primary germ layers of an embryo.

endodermis – a plant tissue composed of a single layer of cells that surrounds and regulates the passage of materials into the vascular cylinder of roots.

endometrium – the inner lining of the uterus.

endoskeleton – hardened, supportive internal tissue of echinoderms and vertebrates.

endosperm – a plant tissue of angiosperm seeds that stores nutrients.

endothelium – the layer of epithelial tissue that forms the thin inner lining of blood vessels and heart chambers.

enzyme – a protein catalyst that activates a specific reaction.

eosinophil – a type of white blood cell that becomes stained by acidic eosin dye; constitutes about 2%–4% of the human white blood cells.

epicardium – the thin, outer layer of the heart; also called the *visceral pericardium*.

epicotyl – part of the plant embryo that contributes to stem development.

epidermis – the outermost layer of the skin, composed of stratified squamous epithelium; also the outer part of plants.

epididymis – a coiled tube located along the posterior border of the testis; stores spermatozoa and discharges them during ejaculation.

epidural space – a space between the spinal dura mater and the bone of the vertebral canal.

epiglottis – a cartilaginous leaf-like structure positioned on top of the larynx that covers the glottis during swallowing.

epinephrine – a hormone secreted from the adrenal medulla resulting in actions similar to those from sympathetic nervous system stimulation; also called *adrenaline*.

epiphyseal plate – a cartilaginous layer located between the epiphysis and diaphysis of a long bone, which functions in longitudinal bone growth.

epiphysis – the end segment of a long bone, distinct in early life but later becoming part of the larger bone.

epiphyte – nonparasitic plant, such as orchid and Spanish moss, that grows on the surface of other plants.

epithelial tissue – one of the four basic tissue types; the type of tissue that covers or lines all exposed body surfaces.

erection – a response within an organ, such as the penis, when it becomes turgid and erect as opposed to being flaccid.

erythrocyte – a red blood cell.

esophagus – a tubular organ of the gastrointestinal tract that leads from the pharynx to the stomach.

estrogen – female sex hormone secreted from the ovarian (Graafian) follicle.

estuary – a zone of mixing between fresh water and sea water.

eukaryotic – possessing the membranous organelles characteristic of complex cells.

eustachian canal – see *auditory tube*.

evolution – genetic and phenotypic changes occurring in populations of organisms through time, generally resulting in increased adaptation for continued survival.

excretion – discharging waste material.

exocrine gland – a gland that secretes its product to an epithelial surface, directly or through ducts.

exoskeleton – an outer, hardened supporting structure secreted by ectoderm or epidermis.

expiration – the process of expelling air from the lungs through breathing out; also called *exhalation*.

extension – a movement that increases the angle between two bones of a joint.

external ear – the outer portion of the ear, consisting of the auricle (pinna), external auditory canal, and tympanum.

extracellular – outside a cell or cells.

extraembryonic membranes – membranes that are not a part of the embryo but are essential for the health and development of the organism.

extrinsic – pertaining to an outside or external origin.

facet – a small, smooth surface of a bone where articulation occurs.

facilitated transport – transfer of a particle into or out of a cell along a concentration gradient by a process requiring a carrier.

fallopian tube – see *uterine tube*.

fascia – a tough sheet of fibrous connective tissue binding the skin to underlying muscles or supporting and separating muscle.

fasciculus – a bundle of muscle or nerve fibers.

feces – waste material expelled from the gastrointestinal tract during defecation, composed of food residue, bacteria, and secretions; also called *stool*.

fertilization – the fusion of two haploid gamete nuclei to form a diploid zygote nucleus.

fetus – the unborn offspring during the last stage of prenatal development.

fibrous root – an intertwining mass of roots of about equal size.

filament – a long chain of cells.

filter feeder – an animal that obtains food by straining it from the water.

filtration – the passage of a liquid through a filter or a membrane.

fimbriae – fringe-like extensions from the borders of the open end of the uterine tube.

fissure – a groove or narrow cleft that separates two parts of an organ.

flagella – long, slender locomotor processes characteristic of flagellate protozoans, certain bacteria, and sperm.

flexion – a movement that decreases the angle between two bones of a joint; opposite of extension.

flora – a general term for the plant life of a region or area.

flower – the blossom of an angiosperm that contains the reproductive organs.

fluke – a parasitic flatworm within the class Trematoda.

follicle – the portion of the ovary that produces the egg and the female sex hormone estrogen; the depression that supports and develops a feather or hair.

fontanel – a membranous-covered region on the skull of a fetus or baby where ossification has not yet occurred; also called a *soft spot*.

food web – the food links between populations in a community.

foot – the terminal portion of the lower extremity, consisting of the tarsus, metatarsus, and digits.

foramen – an opening in a bone for the passage of a blood vessel or a nerve.

foramen ovale – the opening through the interatrial septum of the fetal heart.

fossa – a depressed area, usually on a bone.

fossil – any preserved remains or impressions of an organism.

fourth ventricle – a cavity within the brain containing cerebrospinal fluid.

fovea centralis – a depression on the macula lutea of the eye where only cones are located, which is the area of keenest vision.

frond – a leaf of a fern usually containing many leaflets.

fruit – a mature ovary enclosing a seed or seeds.

gallbladder – a pouchlike organ, attached to the inferior side of the liver, which stores and concentrates bile.

gamete – a haploid sex cell, sperm or egg.

gametophyte – the haploid, gamete-producing generation in the life cycle of a plant.

gamma globulins – protein substances that act as antibodies often found in immune serums.

ganglion – an aggregation of nerve cell bodies outside the central nervous system.

gastrointestinal tract – the tubular portion of the digestive system that includes the stomach and the small and large intestines; *(GI tract)*.

gene – part of the DNA molecule located in a definite position on a certain chromosome and coding for a specific product.

gene pool – the total of all the genes of the individuals in a population.

genetic drift – evolution by chance process.

genetics – the study of heredity.

genotype – the genetic make-up of an organism.

genus – the taxonomic category above species and below family.

geotropism – plant growth oriented with respect to gravity; stems grow upward, roots grow downward.

germ cells – gametes or the cells that give rise to gametes or other cells.

germination – the process by which a spore or seed ends dormancy and resumes normal metabolism, development, and growth.

gill – a gas-exchange organ characteristic of fishes and other aquatic or semiaquatic animals.

gingiva – the fleshy covering over the mandible and maxilla through which the teeth protrude within the mouth; also called the *gum*.

girdling – removal of a strip of bark from around a tree down to the wood layer.

gland – an organ that produces a specific substance or secretion.

glans penis – the enlarged, distal end of the penis.

glomerular capsule – the double-walled proximal portion of a renal tubule that encloses the glomerulus of a *nephron;* also called *Bowman's capsule.*

glomerulus – a coiled tuft of capillaries that is surrounded by the glomerular capsule and filters urine from the blood.

glottis – a slit-like opening into the larynx, positioned between the true vocal folds.

glycogen – the principal storage carbohydrate in animals. It is stored primarily in the liver and is made available as glucose when needed by the body cells.

goblet cell – a unicellular gland within columnar epithelia that secretes mucus.

gonad – a reproductive organ, testis or ovary, that produces gametes and sex hormones.

granum – a "stack" of membrane flattened disks within the chloroplast that contain chlorophyll.

gray matter – the portion of the central nervous system that is composed of nonmyelinated nervous tissue.

grazer – animal that feeds on low-growing vegetation, such as grasses.

growth ring – a growth layer of secondary xylem (wood) or secondary phloem in gymnosperms or angiosperms.

guard cell – an epidermal cell to the side of a leaf stoma that helps to control the stoma size.

gut – pertaining to the intestine; generally a developmental term.

gymnosperm – a vascular seed-producing plant that does not produce flowers.

gyrus – a convoluted elevation or ridge.

habitat – the ecological abode of a plant or animal species.

hair – an epidermal structure consisting of keratinized dead cells that have been pushed up from a dividing basal layer.

hair cells – specialized receptor nerve endings for responding to sensations, such as in the spiral organ of the inner ear.

hair follicle – a tubular depression in the skin in which a hair develops.

hand – the terminal portion of the upper extremity, consisting of the carpus, metacarpus, and digits.

haploid – having one copy of each different chromosome.

hard palate – the bony partition between the oral and nasal cavities, formed by the maxillae and palatine bones.

Haversian system – see *osteon.*

heart – a muscular, pumping organ positioned in the thoracic cavity.

hematocrit – the volume percentage of red blood cells in whole blood.

hemoglobin – the pigment of red blood cells that transports O_2 and CO_2.

hemopoiesis – production of red blood cells.

hepatic portal circulation – the return of venous blood from the digestive organs and spleen through a capillary network within the liver before draining into the heart.

herbaceous – a nonwoody plant.

herbaceous stem – stem of a nonwoody plant.

herbivore – an organism that feeds exclusively on plants.

heredity – the transmission of certain characteristics, or traits, from parents to offspring, via the genes.

heterodont – having teeth differentiated into incisors, canines, premolars, and molars for specific functions.

heterotroph – an organism that utilizes preformed food.

heterozygous – having two different alleles (i.e., *Bb*) for a given trait.

hiatus – an opening or fissure.

hilum – a concave or depressed area where vessels or nerves enter or exit an organ.

histology – microscopic anatomy of the structure and function of tissues.

holdfast – basal extension of a multicellular alga that attaches it to a solid object.

homeostasis – a consistency and uniformity of the internal body environment that maintains normal body function.

homologous – similar in developmental origin and sharing a common ancestry.

homothallic – species in which individuals produce both male and female reproductive structures and are self-fertile.

hormone – a chemical substance that is produced in an endocrine gland and secreted into the bloodstream to cause an effect in a specific target organ.

host – an organism on or in which another organism lives.

ileum – the terminal portion of the small intestine between the jejunum and cecum.

imprinting – a type of learned behavior during a limited critical period.

indigenous – organisms that are native to a specific region; not introduced.

inguinal – pertaining to the groin region.

insertion – the more movable attachment of a muscle, usually more distal in location.

inspiration – the act of breathing air into the pulmonary alveoli of the lungs; also called *inhalation.*

instar – stage of insect or other arthropod development between molts.

integument – pertaining to the skin.

internal ear – the innermost portion or chamber of the ear, containing the cochlea and the vestibular organs.

internode – region between stem nodes.

interstitial – pertaining to spaces or structures between the functioning active tissue of any organ.

intracellular – within the cell itself.

intervertebral disk – a pad of fibrocartilage between the bodies of adjacent vertebrae.

intestinal gland – a simple tubular digestive gland that opens onto the surface of the intestinal mucosa and secretes digestive enzymes; also called *crypt of Lieberkühn.*

intrinsic – situated or pertaining to internal origin.

invertebrate – an animal that lacks a vertebral column.

iris – the pigmented vascular tunic portion of the eye that surrounds the pupil and regulates its diameter.

islets of Langerhans – see *pancreatic islets.*

isotope – a chemical element that has the same atomic number as another but a different atomic weight.

jejunum – the middle portion of the small intestine, located between the duodenum and the ileum.

joint capsule – a fibrous tissue cuff surrounding a movable joint.

jugular – pertaining to the veins of the neck that drain the areas supplied by the carotid arteries.

karyotype – the arrangement of chromosomes that is characteristic of the species or of a certain individual.

keratin – an insoluble protein present in the epidermis and in epidermal derivatives such as scales, feathers, hair, and nails.

kidney – one of the paired organs of the urinary system that contains nephrons and filters wastes from the blood in the formation of the urine.

kingdom – a taxonomic category grouping related phyla.

labia major – a portion of the external genitalia of a female, consisting of two longitudinal folds of skin extending downward and backward from the mons pubis.

labia minora – two small folds of skin, devoid of hair and sweat glands, lying between the labia majora of the external genitalia of a female.

lacrimal gland – a tear-secreting gland, located on the superior lateral portion of the eyeball underneath the upper eyelid.

lacteal – a small lymphatic duct within a villus of the small intestine.

lacuna – a hollow chamber that houses an osteocyte in mature bone tissue or a chondrocyte in cartilage tissue.

lamella – a concentric ring of matrix surrounding the central canal in an osteon of mature bone tissue.

large intestine – the last major portion of the gastrointestinal tract, consisting of the cecum, colon, rectum, and anal canal.

larva – an immature, developmental stage that is quite different from the adult.

larynx – the structure located between the pharynx and trachea that houses the vocal cords; commonly called the *voice box*.

lateral root – a secondary root that arises by branching from an older root.

leaf veins – plant structures that contain the vascular tissues in a leaf.

legume – a member of the pea, or bean, family.

lens – a transparent refractive structure of the eye, derived from ectoderm and positioned posterior to the pupil and iris.

lenticel – spongy area in the bark of a stem or root that permits interchange of gases between internal tissues and the atmosphere.

leukocyte – a white blood cell; also spelled *leucocyte*.

lichen – alga or bacteria and fungi coexisting in a mutualistic relationship.

ligament – a fibrous band of connective tissue that binds bone to bone to strengthen and provide support to the joint; also may support viscera.

limbic system – portion of the brain concerned with emotions and autonomic activity.

locus – the specific location or site of a gene within the chromosome.

lumbar – pertaining to the region of the loins.

lumen – the space within a tubular structure through which a substance passes.

lung – one of the two major organs of respiration within the thoracic cavity.

lymph – a clear fluid that flows through lymphatic vessels.

lymph node – a small, oval mass located along the course of lymph vessels.

lymphocyte – a type of white blood cell characterized by a granular cytoplasm.

macula lutea – a depression in the retina that contains the fovea centralis, the area of keenest vision.

malnutrition – any abnormal assimilation of food; receiving insufficient nutrients.

mammary gland – the gland of the mammalian female breast responsible for lactation and nourishment of the young.

mantle – fleshy fold of the body wall of a mollusk, typically involved in shell formation.

marine – pertaining to the sea or ocean.

marrow – the soft vascular tissue that occupies the inner cavity of certain bones and produces blood cells.

matrix – the intercellular substance of a tissue.

mediastinum – the partition in the center of the thorax between the two pleural cavities.

medulla – the center portion of an organ.

medulla oblongata – a portion of the brainstem between the pons and the spinal cord.

medullary cavity – the hollow center of the diaphysis of a long bone, occupied by red bone marrow.

megaspore – a plant spore that will germinate to become a female gametophyte.

meiosis – cell division by which gametes, or haploid sex cells, are formed. In plants, meiosis yields spores.

melanocyte – a pigment-producing cell in the deepest epidermal layer of the skin.

membranous bone – bone that forms from membranous connective tissue rather than from cartilage.

menarche – the first menstrual discharge.

meninges – a group of three fibrous membranes that cover the central nervous system.

meniscus – wedge-shaped cartilage in certain synovial joints.

menopause – the cessation of menstrual periods in the human female.

menses – the monthly flow of blood from the human female genital tract.

menstrual cycle – the rhythmic female reproductive cycle, characterized by changes in hormone levels and physical changes in the uterine lining.

menstruation – the discharge of blood and tissue from the uterus at the end of the menstrual cycle.

meristem tissue – undifferentiated plant tissue that is capable of dividing and producing new cells.

mesentery – a fold of peritoneal membrane that attaches an abdominal organ to the abdominal wall.

mesoderm – the middle one of the three primary germ layers.

mesophyll – the middle tissue layer of a leaf containing cells that are active in photosynthesis.

mesothelium – a simple squamous epithelial tissue that lines body cavities and covers visceral organs; also called *serosa*.

metabolism – the chemical changes that occur within a cell.

metacarpus – the region of the hand between the wrist and the phalanges, including the five bones that comprise the palm of the hand.

metamorphosis – change in morphologic form, such as when an insect larva develops into the adult or as a tadpole develops into an adult frog.

metatarsus – the region of the foot between the ankle and the phalanges, consisting of five bones.

microbiology – the science dealing with microscopic organisms, including bacteria, fungi, protozoa, and viruses.

microspore – a spore in seed plants that develops into a pollen grain, the male gametophyte.

microvilli – microscopic, hair-like projections of cell membranes on certain epithelial cells.

midbrain – the portion of the brain between the pons and the forebrain.

middle ear – the middle of the three ear chambers, containing the three auditory ossicles.

migration – movement of organisms from one geographical site to another.

mimicry – a protective resemblance of an organism to another.

mitosis – the process of cell division, in which the two daughter cells are identical and contain the same number of chromosomes.

mitral valve – the left atrioventricular heart valve; also called the *bicuspid valve*.

mixed nerve – a nerve containing both motor and sensory nerve fibers.

molecule – a minute mass of matter, composed of a combination of atoms that form a given chemical substance or compound.

molting – periodic shedding of an epidermal derived structure.

monocot – a type of angiosperm in which the seed has only a single cotyledon; also called *monocotyledon*.

motor neuron – a nerve cell that conducts action potential away from the central nervous system and innervates effector organs (muscles and glands); also called *efferent neuron*.

motor unit – a single motor neuron and the muscle fibers it innervates.

mucosa – a mucous membrane that lines cavities and tracts opening to the exterior.

muscle – an organ adapted to contract; three types of muscle tissue are cardiac, smooth, and skeletal.

mutation – a variation in an inheritable characteristic, a permanent transmissible change in which the offspring differ from the parents.

mutualism – a beneficial relationship between two organisms of different species.

myelin – a lipoprotein material that forms a sheath-like covering around nerve fibers.

myocardium – the cardiac muscle layer of the heart.

myofibril – a bundle of contractile fibers within muscle cells.

myoneural junction – the site of contact between an axon of a motor neuron and a muscle fiber.

myosin – a thick filament protein that, together with actin, causes muscle contraction.

nail – a hardened, keratinized plate that develops from the epidermis and forms a protective covering on the surfaces of the digits.

nares – the openings into the nasal cavity; also called *nostrils*.

nasal cavity – a mucosa-lined space above the oral cavity, which is divided by a nasal septum and is the first chamber of the respiratory system.

nasal septum – a bony and cartilaginous partition that separates the nasal cavity into two portions.

natural selection – the evolutionary mechanism by which better adapted organisms are favored to reproduce and pass on their genes to the next generation.

nephron – the functional unit of the kidney, consisting of a glomerulus, glomerular capsule, convoluted tubules, and the nephron loop.

nerve – a bundle of nerve fibers outside the central nervous system.

neurofibril node – a gap in the myelin sheath of a nerve fiber; also called the *node of Ranvier*.

neuroglia – specialized supportive cells of the central nervous system.

neurolemmocyte – a specialized neuroglial cell that surrounds an axon fiber of a peripheral nerve and forms the neurilemmal sheath; also called the *Schwann cell*.

neuron – the structural and functional unit of the nervous system, composed of a cell body, dendrites, and an axon; also called a *nerve cell*.

neutron – a subatomic particle in the nucleus of an atom that has a weight of one atomic mass unit and carries no charge.

neutrophil – a type of phagocytic white blood cell.

niche – the position and functional role of an organism in its ecosystem.

nipple – a dark pigmented, rounded projection at the tip of the breast.

nitrogen fixation – a process carried out by certain organisms, such as by soil bacteria, whereby free atmospheric nitrogen is converted into ammonia or nitrate compounds.

node – location on a stem where a leaf is attached.

node of Ranvier – see *neurofibril node*.

notochord – a flexible rod of tissue that extends the length of the back of an embryo and in some adults.

nucleic acid – an organic molecule composed of joined nucleotides, such as RNA and DNA.

nucleus – a spheroid body within a eukaryotic cell that contains the chromosomes of the cell.

nut – a hardened and dry single-seeded fruit.

olfactory – pertaining to the sense of smell.

oocyte – a developing egg cell.

oogenesis – the process of female gamete formation.

oogonium – a unicellular female reproductive organ of various protists that contains a single or several eggs.

optic – pertaining to the eye and the sense of vision.

optic chiasma – an X-shaped structure on the inferior aspect of the brain where there is a partial crossing over of fibers in the optic nerves.

optic disk – a small region of the retina where the fibers of the ganglion neurons exit from the eyeball to form the optic nerve; also called the *blind spot*.

oral – pertaining to the mouth; also called *buccal*.

organ – a structure consisting of two or more tissues, which performs a specific function.

organelle – a minute structure of the eukaryotic cell that performs a specific function.

organism – an individual living creature.

orifice – an opening into a body cavity or tube.

origin – the place of muscle attachment onto the more stationary point or proximal bone; opposite the insertion.

osmosis – the diffusion of water from a solution of lesser concentration to one of greater concentration through a semipermeable membrane.

ossicle – one of the three bones of the middle ear.

osteocyte – a mature bone cell.

osteon – a group of osteocytes and concentric lamellae surrounding a central canal within bone tissue; also called a *Haversian system*.

oval window – see *vestibular window*

ovarian follicle – a developing ovum and its surrounding epithelial cells.

ovary – the female gonad in which ova and certain sexual hormones are produced.

oviduct – the tube that transports ova from the ovary to the uterus; also called the *uterine tube* or *fallopian tube*.

ovipositor – a structure at the posterior end of the abdomen in many female insects for laying eggs.

ovulation – the rupture of an ovarian follicle with the release of an ovum.

ovule – the female reproductive organ in a seed plant that contains megasporangium where meiosis occurs and the female gametophyte is produced.

ovum – a secondary oocyte after ovulation but before fertilization.

palisade layer – the upper layer of the mesophyll of a leaf that carries out photosynthesis.

pancreas – organ in the abdominal cavity that secretes gastric juices into the gastrointestinal tract and insulin and glucagon into the blood.

pancreatic islets – a cluster of cells within the pancreas that forms the endocrine portion of the pancreas; also called *islets of Langerhans*.

papillae – small nipple-like projections.

paranasal sinus – a mucous-lined air chamber that communicates with the nasal cavity.

parasite – an organism that resides in or on another from which it derives sustenance.

parasympathetic – pertaining to the division of the autonomic nervous system concerned with activities that are antagonistic to the sympathetic division of the autonomic nervous system.

parathyroids – small endocrine glands that are embedded on the posterior surface of the thyroid glands and are concerned with calcium metabolism.

parenchyma – the principal structural cells of plants.

parietal – pertaining to a wall of an organ or cavity.

parotid gland – one of the paired salivary glands on the sides of the face over the masseter muscle.

parturition – the process of childbirth.

pathogen – any disease-producing organism.

pectin – an organic compound in the intercellular layer and primary wall of plant cell walls; the basis of fruit jellies.

pedicel – the stalk of a flower in an inflorescence.

pectoral girdle – the portion of the skeleton that supports the upper extremities.

pelvic – pertaining to the pelvis.

pelvic girdle – the portion of the skeleton to which the lower extremities are attached.

penis – the external male genital organ, through which urine passes during urination and which transports semen to the female during coitus.

perennial – a plant that lives throughout several to many growing seasons.

pericardium – a protective serous membrane that surrounds the heart.

pericarp – the fruit wall that forms from the wall of a mature ovary.

perineum – the floor of the pelvis.

periosteum – a fibrous connective tissue covering the surface of bone.

peripheral nervous system – the nerves and ganglia of the nervous system that lie outside of the brain and spinal cord.

peristalsis – rhythmic contractions of smooth muscle in the walls of various tubular organs, which move the contents along.

peritoneum – the serous membrane that lines the abdominal cavity and covers the abdominal viscera.

petal – the leaf of a flower, which is generally colored.

petiole – structure of a leaf that connects the blade to the stem.

phagocyte – any cell that engulfs other cells, including bacteria, or small foreign particles.

phalanx – a bone of the finger or toe.

pharynx – the region of the gastrointestinal tract and respiratory system located at the back of the oral and nasal cavities and extending to the larynx anteriorly and the esophagus posteriorly; also called the *throat*.

phenotype – the appearance of an organism caused by the genotype and environmental influences.

pheromone – a chemical secreted by one organism that influences the behavior of another.

phloem – vascular tissue in plants that transports nutrients.

photoperiodism – the response of an organism to periods of light and dark.

photosynthesis – the process of using the energy of the sun to make carbohydrate from carbon dioxide and water.

phototropism – plant growth or movement in response to a directional light source.

physiology – the science that deals with the study of body functions.

phytoplankton – microscopic, free-floating, photosynthetic organisms that are the major primary producers in freshwater and marine ecosystems.

pia mater – the innermost meninx that is in direct contact with the brain and spinal cord.

pineal gland – a small cone-shaped gland located in the roof of the third ventricle of the brain.

pistil – a reproductive structure of a flower composed of the stigma, style, and ovary.

pith – a centrally located tissue within a dicot stem.

pituitary gland – a small, pea-shaped endocrine gland situated on the inferior surface of the brain that secretes a number of hormones; also called the *hypophysis*.

placenta – the organ of metabolic exchange between the mother and the fetus.

plankton – aquatic, free-floating microscopic organisms.

plasma – the fluid, extracellular portion of circulating blood.

plastid – an organelle in the cell of certain plants that is the site for food manufacture and storage.

platelets – fragments of specific bone marrow cells that function in blood coagulation; also called *thrombocytes*.

pleural membranes – serous membranes that surround the lungs and line the thoracic cavity.

plexus – a network of interlaced nerves or vessels.

pollen grain – immature male gametophyte generation of seed plants.

pollination – the delivery by wind, water, or animals of pollen to the ovule of a seed plant, leading to fertilization.

polypeptide – a molecule of many amino acids linked by peptide bonds.

pons – the portion of the brainstem just above the medulla oblongata and anterior to the cerebellum.

population – all the organisms of the same species in a particular location.

posterior (*dorsal*) – toward the back.

predation – the consumption of one organism by another.

pregnancy – a condition in which a female has a developing offspring in the uterus.

prenatal – the period of offspring development during pregnancy; before birth.

prey – organisms that are food for a predator.

producer – organisms within an ecosystem that synthesize organic compounds from inorganic constituents.

prokaryote – organism, such as a bacterium, that lacks the specialized organelles characteristic of complex cells.

proprioceptor – a sensory nerve ending that responds to changes in tension in a muscle or tendon.

prostate – a walnut-shaped gland surrounding the male urethra just below the urinary bladder that secretes an additive to seminal fluid during ejaculation.

protein – a macromolecule composed of one or several polypeptides.

prothallus – a heart-shaped structure that is the gametophyte generation of a fern.

proton – a subatomic particle of the atom nucleus that has a weight of one atomic mass unit and carries a positive charge; also called a *hydrogen ion*.

proximal – closer to the midline of the body or origin of an appendage; opposite of distal.

puberty – the period of human development in which the reproductive organs become functional.

pulmonary – pertaining to the lungs.

pupil – the opening through the iris that permits light to pass through the lens and enter the posterior cavity.

radial symmetry – symmetry around a central axis so that any half of an organism is identical to the other.

receptacle – the tip of the axis of a flower stalk that bears the floral organs.

receptor – a sense organ or a specialized end of a sensory neuron that receives stimuli from the environment.

rectum – the terminal portion of the gastrointestinal tract, between the sigmoid colon and the anal canal.

reflex arc – the basic conduction pathway through the nervous system, consisting of a sensory neuron, an interneuron (association), and a motor neuron.

regeneration – regrowth of tissue or the formation of a complete organism from a portion.

renal – pertaining to the kidney.

renal corpuscle – the portion of the nephron consisting of the glomerulus and a glomerular capsule.

renal pelvis – the inner cavity of the kidney formed by the expanded ureter and into which the calyces open.

renewable resource – a commodity that is not used up because it is continually produced in the environment.

replication – the process of producing a duplicate; a copying or duplication, such as DNA replication.

respiration – the exchange of gases between the external environment and the cells of an organism; the metabolic activity of cells resulting in the production of ATP.

retina – the inner layer of the eye that contains the rods and cones.

rhizome – an underground stem in some plants that stores photosynthetic products and gives rise to roots and above-ground stems and leaves.

rod – a photoreceptor in the retina of the eye that is specialized for colorless, dim-light vision.

root – the anchoring subterranean portion of a plant that permits absorption and conduction of water and minerals.

root cap – end mass of parenchyma cells that protects the apical meristem of a root.

root hair – unicellular epidermal projection from the root of a plant, which functions in absorption.

rugae – the folds or ridges of the mucosa of an organ.

sagittal – a vertical plane through the body that divides it into right and left portions.

salinity – saltiness in water or soil; a measure of the concentration of dissolved salts.

salivary gland – an accessory digestive gland that secretes saliva into the oral cavity.

sarcolemma – the cell membrane of a muscle fiber.

sarcomere – the portion of a skeletal muscle fiber between the two adjacent Z lines that is considered the functional unit of a myofibril.

savanna – open grassland with scattered trees.

Schwann cell – see *neurolemmocyte*.

sclera – the outer white layer of connective tissue that forms the protective covering of the eye.

sclerenchyma – supporting tissue in plants composed of hollow cells with thickened walls.

scolex – head region of a tapeworm.

scrotum – a pouch of skin that contains the testes and their accessory organs.

sebaceous gland – an exocrine gland of the skin that secretes *sebum*, an oily protective product.

secondary growth – plant growth in girth from secondary or lateral meristems.

seed – a plant embryo with a food reserve that is enclosed in a protective seed coat; seeds develop from matured ovules.

semen – the secretion of the reproductive organs of the male, consisting of spermatozoa and additives.

semilunar valve – crescent-shaped heart valves, positioned at the entrances to the aorta and the pulmonary trunk.

sensory neuron – a nerve cell that conducts an impulse from a receptor organ to the central nervous system; also called *afferent neuron*.

sepal – outermost whorl of flower structures beneath the petals; collectively called the *calyx*.

serous membrane – an epithelial and connective tissue membrane that lines body cavities and covers viscera; also called *serosa*.

sesamoid bone – a membranous bone formed in a tendon in response to joint stress.

sessile – organisms that lack locomotion and remain stationary, such as sponges and plants.

shoot – portion of a vascular plant that includes a stem with its branches and leaves.

sinoatrial node (SA node) – a mass of cardiac tissue in the wall of the right atrium that initiates the cardiac cycle; also called the *pacemaker*.

sinus – a cavity or hollow space within a body organ such as a bone.

skeletal muscle – a type of muscle tissue that is multinucleated, occurs in bundles, has crossbands of proteins, and contracts either in a voluntary or involuntary fashion.

small intestine – the portion of the gastrointestinal tract between the stomach and the cecum; functions in absorption of food nutrients.

smooth muscle – a type of muscle tissue that is nonstriated, composed of fusiform, single-nucleated fibers, and contracts in an involuntary, rhythmic fashion within the walls of visceral organs.

solute – a substance dissolved in a solvent to form a solution.

solvent – a fluid such as water that dissolves solutes.

somatic – pertaining to the nonvisceral parts of the body.

somatic cells – all the cells of the body of an organism except the germ cells.

sorus – a cluster of sporangia on the underside of fern pinnae.

species – a group of morphologically similar (common gene pool) organisms that are capable of interbreeding and producing fertile offspring and reproductively isolated.

spermatic cord – the structure of the male reproductive system composed of the ductus deferens, spermatic vessels, nerve, cremasteric muscle, and connective tissue.

spermatogenesis – the production of male sex gametes, or spermatozoa.

spermatozoon – a sperm cell, or gamete.

sphincter – a circular muscle that constricts a body opening or the lumen of a tubular structure.

spinal cord – the portion of the central nervous system that extends from the brainstem through the vertebral canal.

spinal nerve – one of the 31 pairs of nerves that arise from the spinal cord.

spiracle – a respiratory opening in certain animals such as arthropods and sharks.

spirillum (pl. spirilla) – a spiral-shaped bacterium.

spleen – a large, blood-filled organ located in the upper left of the abdomen and attached by the mesenteries to the stomach.

spongy bone – a type of bone that contains many porous spaces; also called *cancellous bone*.

sporangium – any structure within which spores are produced.

spore – a reproductive cell capable of developing into an adult organism without fusion with another cell.

sporophyll – modified leaves that bear one or more sporangia.

stamen – a reproductive structure of a flower, composed of a filament and an anther, where pollen grains are produced.

starch – carbohydrate molecule synthesized from photosynthetic products; common food storage substance in many plants.

stele – the vascular tissue and pith or ground tissue at the central core of a root or stem.

stigma – the upper portion of the pistil of a flower.

stoma – an opening in a plant leaf through which gas exchange takes place.

stomach – a pouch-like digestive organ between the esophagus and the duodenum.

style – the long slender portion of the pistil of a flower.

submucosa – a layer of supportive connective tissue that underlies a mucous membrane.

succession – the sequence of ecological stages by which a particular biotic community gradually changes until there is a community of climax vegetation.

sucrose – a disaccharide (double sugar) consisting of a linked glucose and fructose molecule; the principal transport sugar in plants.

surfactant – a substance produced by the lungs that decreases the surface tension within the pulmonary alveoli.

suture – a type of immovable joint articulating between bones of the skull.

symbiosis – a close association between two organisms in which one or both species derive benefit.

sympathetic – pertaining to that part of the autonomic nervous system concerned with activites antagonistic to the parasympathetic.

synapse – a minute space between the axon terminal of a presynaptic neuron and a dendrite of a postsynaptic neuron.

syngamy – union of gametes in sexual reproduction; fertilization.

synovial cavity – a space between the two bones of a synovial joint, filled with synovial fluid.

system – a group of body organs that function together.

systole – the muscular contraction of the ventricles of the heart during the cardiac cycle.

systolic pressure – arterial blood pressure during the ventricular systolic phase of the cardiac cycle.

taproot – a plant root system in which a single root is thick and straight.

target organ – the specific body organ that a particular hormone affects.

tarsus – pertaining to the ankle; the proximal portion of the foot that contains the tarsal bones.

taxonomy – the science of describing, classifying, and naming organisms.

tendo calcaneous – the tendon that attaches the calf muscles to the calcaneous bone.

tendon – a band of dense regular connective tissue that attaches muscle to bone.

testis – the primary reproductive organ of a male, which produces spermatozoa and male sex hormones.

tetrapod – a four-appendaged vertebrate, such as amphibian, reptile, bird, or mammal.

thoracic – pertaining to the chest region.

thorax – the chest.

thymus gland – a bi-lobed lymphoid organ positioned in the upper mediastinum, posterior to the sternum and between the lungs.

tissue – an aggregation of similar cells and their binding substances, joined to perform a specific function.

tongue – a protrusible muscular organ on the floor of the oral cavity.

toxin – a poisonous compound.

trachea – a tubule in the respiratory system of some invertebrates; the airway leading from the larynx to the bronchi in the respiratory system of vertebrates; also called the *windpipe*.

tract – a bundle of nerve fibers within the central nervous system.

trait – a distinguishing feature studied in heredity.

transpiration – the evaporation of water from a leaf, which pulls water from the roots through the stem of a leaf.

tricuspid valve – the heart valve between the right atrium and the right ventricle; also called *right atrioventricular valve*.

turgor pressure – osmotic pressure that provides rigidity to a cell.

tympanic membrane – the membranous eardrum positioned between the outer ear and middle ear; also called the *tympanum*, or the *eardrum*.

umbilical cord – a cord-like structure containing the umbilical arteries and vein, which connects the fetus with the placenta.

umbilicus – the site where the umbilical cord was attached to the fetus; also called the *navel*.

ureter – a tube that transports urine from the kidney to the urinary bladder.

urethra – a tube that transports urine from the urinary bladder to the outside of the body.

urinary bladder – a distensible sac in the pelvic cavity that stores urine.

uterine tube – the tube through which the ovum is transported to the uterus and where fertilization takes place; also called the *oviduct* or *fallopian tube*.

uterus – a hollow, muscular organ in which a fetus develops; located within the female pelvis between the urinary bladder and the rectum.

uvula – a fleshy, pendulous portion of the soft palate that blocks the nasopharynx during swallowing.

vacuole – a fluid-filled organelle.

vagina – a tubular organ that leads from the uterus to the vestibule of the female reproductive tract and receives the male penis during coitus.

vascular tissue – plant tissue composed of xylem and phloem; functions in transport of water, nutrients, and photosynthetic products throughout the plant.

vegetative – plant parts not specialized for reproduction; asexual reproduction.

veins – blood vessels that convey blood toward the heart.

ventral (anterior) – toward the front surface of the body.

vertebrate – an animal that possesses a vertebral column.

vestibular folds – the supporting folds of tissue for the vocal folds within the larynx; also called *vocal cords*.

vestibular window – a membrane-covered opening in the bony wall between the middle and inner ear into which the footplate of the stapes fits; also called *oval window*.

viscera – the organs within the abdominal or thoracic cavities.

vitreous humor – the transparent gel that occupies the space between the lens and retina of the eye.

vocal folds – folds of the mucous membrane in the larynx that produce sound as they are pulled taut and vibrated; also called *vocal cords*.

vulva – the external genitalia of the female that surround the opening of the vagina; also called the *pudendum*.

wood – interior tissue of a tree composed of secondary xylem.

zoospore – a flagellated or ciliated spore produced asexually by some protists.

zygote – a fertilized egg cell formed by the union of a sperm and an ovum.